SECRETS OF THE GREATEST SNOW ON EARTH

SECRETS OF THE GREATEST SNOW ON EARTH

WEATHER, CLIMATE CHANGE, AND FINDING
DEEP POWDER IN UTAH'S WASATCH
MOUNTAINS AND AROUND THE WORLD

JIM STEENBURGH

SECOND EDITION

UTAH STATE UNIVERSITY PRESS

Logan

© 2023 by University Press of Colorado

Published by Utah State University Press
An imprint of University Press of Colorado
1624 Market Street, Suite 226
PMB 39883
Denver, Colorado 80202–1559

 The University Press of Colorado is a proud member of
Association of University Presses.

The University Press of Colorado is a cooperative publishing enterprise supported, in part, by
Adams State University, Colorado State University, Fort Lewis College, Metropolitan State University
of Denver, University of Alaska Fairbanks, University of Colorado, University of Denver, University of Northern Colorado, University of Wyoming, Utah State University, and Western Colorado
University.

∞ This paper meets the requirements of the ANSI/NISO Z39.48-1992 (Permanence of Paper).

ISBN: 978-1-64642-428-3 (paperback)
ISBN: 978-1-64642-429-0 (ebook)
https://doi.org/10.7330/9781646424290

Cataloging-in-Publication data for this title is available online at the Library of Congress

Photograph credits. Front cover, clockwise from top: © Chris Pearson/Ski Utah; courtesy Adam Naisbitt; courtesy Tyler Cruickshank; © Chris Pearson/Ski Utah; Public-domain image from https://worldview.earthdata.nasa.gov/. Back cover: background photo © Chris Pearson/Ski Utah; author photo by Fabien Maussion. Interior: photo by Aaron Burden on Unsplash (pp. i, 74–75); photo by Jim Steenburgh (pp. ii–iii, v, 20–21, 58–59, 160–161); © Chris Pearson/Ski Utah (pp. x–1, 2–3, 6–7, 136–137); photo by christiannafzger/iStock (pp. 42–43); public-domain image from https://worldview.earthdata.nasa.gov (pp. 92–93); photo by Dana Dagle Photography/iStock (pp. 122–123); photo by FashionStock.com/Shutterstock (pp. 174–175).

To my parents, Jack and Janet, for instilling
my passion for mountains and meteorology

To my wife, Andrea, for her enduring love and
tolerance of my alpine and scientific pursuits

To my children, Erik and Maria, for being the
greatest adventure of my life

Contents

Acknowledgments

I am indebted to many individuals and groups who have made my life richer and this book possible through their friendships and contributions. My parents provided the initial direction (or, more accurately, push) down the personal and professional path that led me to skiing and meteorology. My wife, Andrea, and children, Erik and Maria, gave their love, support, and encouragement through long hours and many ups and downs during the writing and publishing process. Cliff Mass and Dave Schultz have served as friends and mentors throughout my career. Larry Dunn, the now-retired meteorologist-in-charge at the National Weather Service Forecast Office in Salt Lake City, took me on my first backcountry ski tour in Utah, and taught me how to be a true powder snob. Tyler Cruickshank served as my best friend and a voice of reason on hundreds of ski tours in the Wasatch backcountry. My former and current research assistants and colleagues in our Mountain Meteorology Group at the University of Utah taught me more about mountain weather and climate than I could ever hope to learn on my own. They include Trevor Alcott, Tom Blazek, Leah Campbell, Marcel Caron, Will Cheng, Kirby Cook, Justin Cox, Julie Cunningham, Ashley Evans, Robert Grandy, Eric Grimit, Tom Gowan, Scott Halvorson, Ken Hart, Sebastian Hoch, John Horel, Matt Jeglum, Wyndam Lewis, Jeff Massey, John McMillan, Dallas McKinney, Colby Neuman, Daryl Onton, Michael Pletcher, Jon Rutz, Jay Shafer, Andy Siffert, Jebb Stewart, Leigh Sturges, Peter Veals, Michael Wasserstein, Greg West, Tyler West, Dave Whiteman, and Kristen Yeager. In 2019, I spent six months as a Fulbright scholar at the University of Innsbruck, which enabled me to develop a deeper understanding of Alpine meteorology and climate change in mountainous regions. I am grateful to the Fulbright Scholar Program, Fulbright Austria, and the University of Innsbruck for making my visit possible and to Alexander Gohm, Manuela Lehner, Fabi Maussion, Michael Kuhn, Lindsey Nicholson, Mathias Rotach, Iva Stiperski, and many others in the Department of Atmospheric and Cryospheric Sciences for showing me the Tyrolean Alps and collaborating on teaching and research endeavors.

Other individuals and groups who provided valuable guidance, assistance, or contributions include Clinton Alden; Alta ski area; the American Meteorological Society; Seth Arens; Ned Bair; Annie Burgess; Joey Camps; Titus Case; Lee Cohen; Tony Crocker and his bestsnow.net website; Erik

Crosman; Lynn and Patrick de Freitas; Laurie Delaney, Liam Fitzgerald, and the Utah Department of Transportation; the Electron and Confocal Microscopy Laboratory (USDA); Susie English, Hailey Klotz, Chris Pearson, Nathan Rafferty, and Ski Utah; ESRI; Cale Fallgatter; GR Fletcher; Howie Garber; Tim Garrett; Bart Geerts; Joel Gratz; Ethan Greene; Steve Griffin and the *Salt Lake Tribune*; Drew Hardesty; Howie Howlett; Jake Hutchinson; the Integrated Assessment Modeling Consortium; Francesco Isotta and MeteoSuisse; Tim Jansa; the Japanese Meteorological Agency; Randy Julander and the Natural Resources Conservation Service; Mike Kok; Mike and Tom Korologos; Neil Lareau; Christian Lackner; Casey Lenhart and Meteorological Solutions, Inc.; Ken Libbrecht; the J. Willard Marriot Library; Connie Marshall; MesoWest; Angelina Miller; Emily Moench; Jonathan Morgan; Mt. Baker ski area; Sento Nakai; Bill Nalli; NASA; the National Climatic Data Center; the NOAA/National Weather Service; Marc Olefs; Heidi Orchard and the Utah State Historical Society; Brandon Ott; Tom Painter; the PRISM Climate Group at Oregon State University; Dave Richards; Dave Schultz; Bob Smith; Snowbird; Jackson Sponaugle; Maria Steenburgh; Court Strong; Carolyn Stwertka; Andrew Tait and NIWA; Evan Thayer; Bruce Tremper; Unidata; the University of Utah Center for High Performance Computing; the USGS; the Utah Avalanche Center; Wendy Wagner; Christy Wall; Roy Webb; the Western Region Climate Center; Doug Wewer; Brittany Whitlam; Onno Wieringa; Dave Williams and the Utah Office of Tourism; WorldClim; Satoru Yamaguchi; David Yorty; and Jamie Yount. I am also grateful to Robin DuBlanc, Laura Furney, Rachael Levay, Dan Pratt, Michael Spooner, Beth Svinarich, and associates at Utah State University Press and University Press of Colorado for making this book a reality.

Adam Naisbitt of the University of Utah Digit Lab skillfully drafted several figures and provided many great photographs for the first edition of this book that are also used in this second edition. Adam's tragic death in 2017 is a reminder of the need for greater mental health awareness in our lives and in the snow community.

Much of this book is the outgrowth of knowledge that I have gained during my tenure at the University of Utah. I thank the Department of Atmospheric Sciences for hiring me as a young assistant professor and the National Science Foundation, National Oceanic and Atmospheric Administration, National Weather Service, NASA, US Forest Service, and Office of Naval Research for financial support that enabled many advances in our understanding of the weather and climate of northern Utah and other mountainous regions.

Introduction

Is it true? Does Utah really have the Greatest Snow on Earth? What about claims that Utah's snow is lighter and drier than elsewhere, that magic snow-flakes are created because the western deserts dry out snow, or that moisture from the Great Salt Lake fuels storms?

The first meteorologist to ponder these questions was S. D. Green in the 1930s. Green was an avid skier who worked for the US Weather Bureau (now the National Weather Service). Lake Placid had just hosted the 1932 Olympics and was a favorite winter-sports destination for easterners. The West, however, was largely unknown to skiers. In an article published in the *Salt Lake Tribune* in 1935, Green argued that the "natural advantages" of Lake Placid were inferior to those of Utah and that upper Big and Little Cottonwood Canyons offered the best skiing in the Wasatch Mountains (Kelner 1980, 155).

In the late 1930s, lifts were installed in the Cottonwood Canyons, and Alta quickly became a mecca for skiers who wanted to avoid the packed slopes of the Alps or the eastern United States. In the 1940s and 1950s, Fred Speyer, Dick Durrance, Sverre Engen, Alf Engen, and Dolores LaChapelle pioneered techniques for deep-powder skiing at Alta. Their new approaches to skiing could have been developed only at a ski area with abundant, high-quality, nat-ural snowfall. Enthusiastically taken up by European ski professionals, these techniques, as noted by Lou Dawson, "spread around the world like pollen in strong wind" (Dawson 1997, 166).

If Utah were to become a powder paradise, however, techniques to minimize avalanche hazard following major storms needed to be developed; in the 1940s, these techniques didn't exist. Shortly after World War II, the US Forest Service appointed Monty Atwater as Alta's snow ranger. Recognizing that adventur-ers flocked to Alta to ski powder, not to be hemmed in by ropes and closed-area signs, Atwater and fellow avalanche hunter Ed LaChapelle pioneered the use of explosives and artillery to intentionally trigger avalanches before they became life-threatening menaces. Alta became the place for training in snow science and avalanche-mitigation techniques.

With a snow reputation firmly established, skiing blossomed in the Wasatch Mountains. Today, winter recreation represents a $1 billion a year industry for the State of Utah. More and more recreationists are venturing into the Wasatch backcountry. The slogan Greatest Snow on Earth is one of the most success-ful in outdoor recreation. For many who come to Utah, powder is more than snow. It is a way of life.

Having grown up in upstate New York, my first western ski experience was in January 1986. The trip was a high school graduation gift from my father. We

skied on long, skinny racing skis, slept in a Motel 6 in the Salt Lake Valley, and hit five ski areas, including Alta, where we had an epic bluebird powder day. Lift tickets at Solitude were $5. As a budding young meteorologist who had just had his first taste of the Greatest Snow on Earth, I, like S. D. Green before me,

Figure 0.1. The author ski touring in the Hida Mountains, Japan. Courtesy Peter Veals.

began to wonder about the "natural advantages" of Big and Little Cottonwood Canyons.

Over the next ten years, I aligned my ski and science passions, eventually earning a PhD in atmospheric sciences from the University of Washington with a specialty in mountain meteorology. Incredibly, I was offered a position at the University of Utah right after graduation. It was 1995 and Salt Lake City had just scored the 2002 Olympic Winter Games. I seized the opportunity and have since spent much of my career studying winter storms and mountain weather around the world (figure 0.1). I've spent as much of my free time as possible skiing.

I've written this book to set the record straight: to dig into Wasatch weather lore, expose the myths, explain the reality, and tell people the real reasons why Utah's powder skiing and snowboarding are so incredible. *Secrets of the Greatest Snow on Earth* is a meteorological guide not only to the weather and climate of the Wasatch Mountains but also to mountain weather and snow around the world. It is written for skiers, snowboarders, weather weenies, and anyone else who can't sleep when the flakes start to fly. I hope it will help you find deep powder and bluebird skies.

1 The Secrets

On December 4, 1960, the legend was born. Inspired by a recent visit of the Ringling Bros. and Barnum & Bailey Circus, a young editor named Tom Korologos opened a special ski edition of the *Salt Lake Tribune's Home Magazine* with the headline "The Greatest Snow on Earth" (figure 1.1). Tom exclaimed, "Intermountain folk will tell you that the winds blowing from the west leave the wet, sticky snows in the Sierras. When the storms reach the Intermountain ranges, only the most perfect dry powder is left. That's just a sprinkling of what you'll find in this vast, scenic country that is the Intermountain area. And what an area. It's some 600 miles long and 2.5 miles high. That's the extent of the Intermountain's big top which supports this real, true Greatest Snow on Earth."

The State of Utah began using Greatest Snow on Earth as a slogan in 1962 and engraved it on license plates in 1985, winning a plate of the year award from the Automobile License Plate Collectors Association (figure 1.2). Utah's trademark on the slogan survived a court challenge from the Ringling Bros. and Barnum & Bailey Circus in the 1990s; the courts ruled that Greatest Snow on Earth doesn't dilute the circus's slogan. But was Tom Korologos right? Is Utah's snow really the greatest on Earth?

No scientist can answer that question. The greatness of snow, like beauty, is in the eye of the beholder. There is no doubt, however, that skiers and snowboarders believe there is something special about Utah snow. Utah ski areas, especially those in Big and Little Cottonwood Canyons southeast of Salt Lake City (figure 1.3), are perennially ranked at or near the top for powder in North America. Alta and Snowbird in upper Little Cottonwood Canyon are frequently co-listed as number one.

What makes the snow in Utah so special? Many people believe, as once proclaimed by Alta Lodge, that "it is a scientific fact that Utah's snow is lighter and drier." Others argue that the snow in Utah is superior because of the Great Salt Lake or the drying influence of upstream deserts. However, these are not the secrets of the Greatest Snow on Earth. It turns out that Utah snow isn't even the lightest and driest in the United States.

Figure 1.2. Classic and modern Greatest Snow on Earth license plates. Source: Zul32, Wikipedia Commons, CC BY-SA 3.0.

SNOW WATER CONTENT

The density of snow depends on its **water content**, the percentage of the snow that is frozen or liquid water. Light, dry snow has a low water content, is fluffy and easy to shovel, and is a breeze to ski through (figure 1.4). Heavy, wet snow has a high water content, is dense and difficult to shovel, and is often more challenging to ski. Meteorologists classify new snow as light when it has a water content of less than 7 percent, average when it has a water content of 7 to 11 percent, and heavy when it has a water content of greater than 11 percent. Artificial snow has a water content of 24 to 28 percent, which is why it is great for base building and not much else. After a storm, freshly fallen snow becomes denser with time as the ice crystals settle and the air spaces shrink. By spring, the water content of a settled natural snowpack is usually between 40 and 50 percent.

Determining the water content of new snow requires measuring the snow depth and the **snow water equivalent**, the depth of the water you would have

Figure 1.3. Geography and major ski areas of the Wasatch Mountains and northern Utah. Big Cottonwood Canyon (BCC) and Little Cottonwood Canyon (LCC) identified with abbreviations. Background hillshade sources: Esri, USGS, FAO, NOAA.

after the snow melts. To measure new snow depth, a white, wooden board is usually placed on the ground or snowpack in a wind-sheltered area prior to a storm (figure 1.5). During or immediately after the storm, snow-depth measurements are made on the board with a ruler or using an **ultrasonic snow-depth sensor**, a clever device that bounces sound waves off the snow surface to measure the snow depth. To obtain the snow water equivalent, one collects and melts down (or weighs) a snow sample from the board. The ratio of the snow water equivalent to the snow depth expressed as a percentage is the water content. For example, if the snow water equivalent is an inch and the snow depth is ten inches, then the water content is $(1/10) \times 100\% = 10\%$.

Figure 1.4. Top: Patti Formichelli enjoys effortless turns through low water content snow in the Wasatch backcountry. © Howie Garber. Used with permission. Bottom: Neil Lareau muscles telemark turns through high water content snow in the Wasatch backcountry.

Figure 1.5. Measuring new snow depth, snow water equivalent, and water content with a snowboard and coring tube. This storm produced 4.8 inches of snow and 0.47 inches of water for a water content of 10.2 percent.

Snow measurement sounds straightforward but is difficult in practice. Snow depth and water content can vary dramatically over short distances, especially if the wind is blowing. Take some snow measurements in your yard during a windy snowstorm and you'll see what I mean. Another issue is the frequency of measurement. Snow settles with time, which decreases the snow depth and increases water content. Four measurements collected every six hours and added will yield a greater total snow depth and lower water content than a single measurement collected at the end of the twenty-four-hour period.

As a result, snow-depth and water-content reports are not precise and are sometimes quite poor. An overenthusiastic ski area can easily game the system by taking frequent observations or sampling in locations that collect the most snow. Even observations collected by meteorologists and **snow-safety profes-sionals** (women and men involved in the study, forecasting, and mitigation of avalanches) are subject to errors and are sparse in coverage. Nevertheless, the data that are available, while not perfect, allow us to evaluate if Utah snow is unusually light and dry.

Snow Water Content in the United States

Observations collected by National Weather Service volunteers show that the average water content of new snow in the United States varies region-ally (figure 1.6). In the western United States, snow with the highest average water content falls in the Pacific states, which include the Cascade Mountains and Sierra Nevada where the snow is known as **Cascade concrete** or **Sierra cement**. These mountain ranges feature a **maritime snow climate** with mild temperatures and heavy snow. Snow with the lowest average water content falls over portions of Montana, Wyoming, and Colorado, where it is often called **cold smoke**, **champagne powder**, or **blower pow**. These areas feature a **continental snow climate** characterized by low temperatures and light snow. Utah lies between these climate regimes and has a **transitional snow climate**.[1] As a result, the snow in Utah has a lower water content than that found in the Pacific states but a slightly higher water content than that found in Montana, Wyoming, and western Colorado. The average water content of Utah snow is also comparable to or slightly higher than that found in the snowbelts near the eastern Great Lakes.

Remember that these are average values and snow water content can vary significant from storm to storm and even during storms. It is possible to ski champagne powder in the Cascades and Sierra or concrete in Utah and

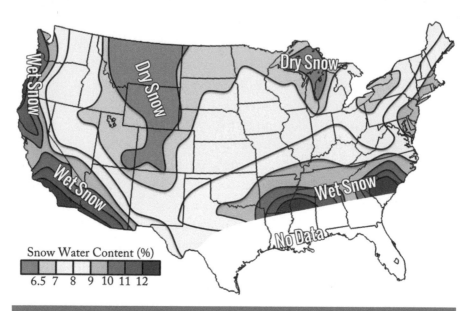

Figure 1.6. Average water content of freshly fallen snow in the continental United States. Values may be lower than observed due to measurement techniques used. Adapted from Baxter et al. 2005.

Colorado. However, a greater fraction of the snow that falls in the Cascades and Sierra has a high water content, whereas a greater fraction that falls over the interior western United States has a low water content.

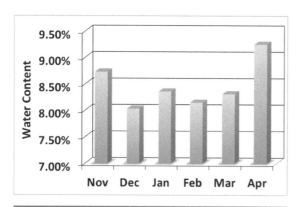

Figure 1.7. Average water content of snow at Alta by month. Source: Steenburgh and Alcott 2008.

Snow Water Content at Alta and Other Mountain Locations

The analysis above suggests that Utah snow is not unusually light and dry, and this is confirmed by snowfall observations collected by snow-safety professionals at Alta ski area, the gold standard for powder skiing in Utah, if not the world. During the ski season, which runs from November through April, the average water content of new snow at Alta is 8.4 percent. By month, the average water content of new snow is slightly lower from December through March and slightly higher in November and April (figure 1.7).

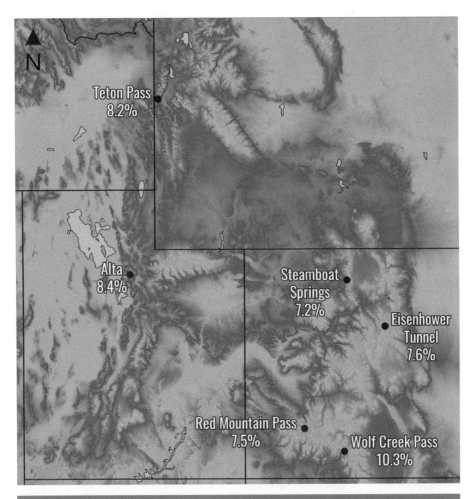

Figure 1.8. Average water content of freshly fallen snow at selected sites in the western United States. Sources: Judson and Doesken 2000; Steenburgh and Alcott 2008. Background hillshade sources: Esri, USGS, FAO, NOAA.

How does this compare to other mountain locations in the western United States? It is certainly lower than that found in the maritime snow climate of the Cascade Mountains and Sierra Nevada. For example, at the Central Sierra Snow Lab in the Sierra Nevada, the average water content of new snow is about 12 percent, which means that Sierra cement is about 1.5 times denser than Utah's Greatest Snow on Earth. On the other hand, the water content of new snow at several mountain sites in the Tetons and the Colorado Rockies is 7.2–8.2 percent, lower than that at Alta (figure 1.8). An exception is the 10.3 percent observed at Wolf Creek Pass in southwest Colorado, where warm, southwesterly flow impinging on the San Juan Mountains frequently produces higher water content snow.

Although there is some uncertainty in these numbers given the difficulties of measuring snow depth and water content, they clearly debunk the myth that Utah snow is the lightest and driest. However, it is also a myth that dry snow produces the best deep-powder skiing.

THE SECRETS OF GREAT POWDER SKIING

Legendary avalanche researcher and powder skier Ed LaChapelle knew more about snow than anyone. While an avalanche hunter at Alta, Ed recognized that "the best deep-powder skiing is not found in the lightest snow, but rather in snow with enough 'body' to provide good flotation for the running ski" (LaChapelle 1962). In other words, there's more to great powder skiing than light, dry snow.

There are three ingredients for great powder skiing. The first is the amount of new snow. This is obvious, but how much is needed? For real powder skiing, the skis or snowboard must float in the new snow. The powder must be **bottomless** so that your skis or snowboard do not ride on the underlying surface. Legendary Taos ski instructor Lito Tejada-Flores once suggested that real powder skiing requires at least a foot of new snow. This is a pretty good estimate, but for our discussion, I'll use ten inches as the minimum snowfall required to qualify as a deep-powder day at a ski area. This lower threshold reflects modern fat skis and snowboards, which float more easily than the narrow skis of yesteryear. However, there is also an upper limit, known as "too much of a good thing." Huge storms create dangerous avalanche conditions, forcing the closure of steep terrain at ski areas, and the deep snow makes it difficult to break trail or maintain momentum when skiing or snowboarding lower-angle slopes in the **backcountry** (figure 1.9a). The best powder skiing comes in **Goldilocks storms:** those that aren't too small or too big, but just right (figure 1.9b).

The second ingredient is a soft underlying surface. Deep-powder skiing is possible with less than ten inches of snow if it falls on settled powder from a previous storm. Such conditions are rare, however, at ski areas, which are usually tracked out each day, but can be found outside the area boundaries in the backcountry. In contrast, ten inches of new snow might not be enough for bottomless skiing if the snow is bone dry and falls on a hard packed or icy snow surface. When such **dust-on-crust** conditions exist (figure 1.9c), the

o Much of a Good Thing   **b.** Goldilocks   **c.** Dust on Crust

Figure 1.9. (a) Six-foot-six meteorologist Matt Jeglum breaks trail through "too much of a good thing" in Killyon Canyon, Utah. Courtesy Dave Marlaire. (b) The author enjoys a Goldilocks storm in the Big Cottonwood backcountry. Courtesy Tyler Cruickshank. (c) Ian Cruickshank samples dust-on-crust in the Big Cottonwood backcountry. Courtesy Tyler Cruickshank.

skiing looks great, but your skis and snowboards sink right through the dry powder and ride on the underlying hardpack.

The third ingredient is a **right-side-up snowfall**, which means that lighter snow sits on top of heavier snow. This **vertical profile of snow water content** is critical for great powder skiing because it helps skis and snowboards float. Right-side-up snow is often called **hero snow** because the skis or snowboard float easily, turns can be made with little or no effort, and skiers and snowboarders feel invincible. On the other hand, in an **upside-down snowfall**, heavy snow sits on top of lighter snow. Skis tend to dive and remain submerged. Fat skis and snowboards help in these conditions but turning is more difficult and requires a more refined technique.

It is the vertical profile of snow water content, not the average water content, that determines the quality of the powder skiing. Imagine two storms, each with an average snow water content of 8 percent. The storm that starts with 12 percent snow and ends with 4 percent produces a right-side-up snowfall and sublime powder skiing and snowboarding conditions. The storm that begins with 4 percent snow and ends with 12 percent produces an upside-down snowfall and difficult skiing and snowboarding conditions.

Knowing these ingredients, we can design the ultimate powder climate. We want abundant, high-quality, natural snowfall with frequent storms that produce at least ten inches of snow. We also want storms that produce right-side-up snow for ski flotation. The snow climate found in the Cottonwood Canyons of Utah's Wasatch Mountains has these ingredients in spades, as illustrated by meteorological records from Alta at the top of Little Cottonwood Canyon.

ALTA'S SNOW CLIMATE

Alta is one of the snowiest ski areas in the world. The average seasonal (November–April) snowfall based on long-term records from the Alta Guard observing site near the base of the ski area is 497 inches. Including snowfall from other months easily pushes the annual average snowfall over 500 inches. Nearby Snowbird, Brighton, and Solitude, which along with Alta comprise the major ski areas in the Cottonwood Canyons, lag only slightly behind (see chapter 2). The snow in the Cottonwood Canyons is not the world's driest, but it is certainly high quality and remarkably consistent. From December through March, the average monthly snowfall at Alta Guard is between 83 and 93 inches (figure 1.10). That equates to an average of over a foot every five days. Snowfall is only slightly lower in the shoulder months of November and April.

Alta averages eighteen days per season (November to April) with at least ten inches of snow, or about one deep-powder day every ten days. This is a mind-boggling number that reflects a high frequency of Goldilocks storms. Most regions that feature drier snow see far fewer deep-powder days. For example, Berthoud Pass, one of the snowier locations in Colorado with an average annual snowfall of almost 400 inches, averages only four days with at least ten inches of snow per season. Alta and the other ski areas

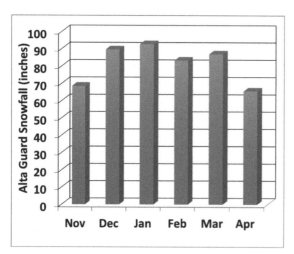

Figure 1.10. Average monthly snowfall at Alta Guard. Sources: Utah Department of Transportation and Utah Avalanche Center.

in the Cottonwood Canyons do see storms that produce "too much of a good thing" or the occasional drought, but the frequency of Goldilocks storms is quite high.

But there is more to the story. At Alta and other Wasatch Mountain ski areas, the climatology favors storms that produce right-side-up snowfalls. In the Wasatch Mountains, snow water content is strongly related to temperature and is typically higher when it is warmer and lower when it is colder. Storms that start warm and get colder with time tend to produce right-side-up snowfalls. Conversely, storms that start cold and get warmer with time tend to produce upside-down snowfalls. On 65 percent of the days that produce ten inches or more of snow at Alta, the temperature at mountaintop level (about 10,000 feet) decreases. On 39 percent of these days, it decreases more than 5°F, whereas on only 18 percent of these days, it increases more than 5°F. In other words, the storm climatology at Alta and in the Cottonwood Canyons is biased to produce lots of right-side-up snowfalls and few upside-down snowfalls.

Thus, the secrets of the Greatest Snow on Earth are abundant natural snowfall, frequent Goldilocks storms, and storm characteristics favoring right-side-up snowfalls. How the climate of Alta compares to the rest of the Wasatch Mountains and other regions of the world is the subject of our next three chapters.

NOTE

1. The boundary between transitional and continental snow climates is not quite coincident with the dry snow region in figure 1.6. For example, the Teton Range of northwest Wyoming and mountain ranges of western Montana have transitional snow climates.

January 22, 1964: Brighton ski area receives 35 inches of snow, Utah's largest observed twenty-four-hour snowfall to date.

January 24–30, 1965: A multiday storm produces 105 inches of snow at Alta.

October–December 1976: One of the worst starts to a ski season in Utah history, with Alta observing only 26.5 inches of snow through the end of December. Even today, locals scornfully refer to the 1976–1977 season as "the drought year."

July 1982–June 1983: Alta observes 847 inches of snow, the second-largest July to June total ever observed in Utah.

December 1983: Alta observes 244.5 inches of snow, establishing the state record for snowfall in a month.

November 22–23, 1992: Alta receives 45 inches of snow in twenty-four hours, establishing a new state record for twenty-four-hour snowfall.

January 4–5, 1994: Alta observes the current state record twenty-four-hour snowfall with 55.5 inches of snow with a water content of 5.7 percent.

November 22–25, 1994: A four-day storm with periods of lake effect produces 92 inches of snow at Alta and 64 inches at Snowbird.

January 14–16, 1995: A three-day storm dumps 100 inches of snow on Alta in sixty-eight hours.

November 22–27, 2001: The "Hundred Inch Storm" produces 108 inches of snow at Alta, including 100 inches in 100 hours over Thanksgiving weekend. With only 10 inches of snow on the ground before the storm, Alta wasn't open prior to Thanksgiving and couldn't open during the holiday weekend due to "too much of a good thing."

October 2004: A remarkable October brings more than 100 inches of snow to the upper Cottonwood Canyons. The Utah Avalanche Center issues its earliest advisory ever on October 20, and Brighton opens on October 29 with a 59-inch base.

May 17–20, 2011: The average snowfall in May is only 27 inches, but this late-season storm dumped 37 inches on Alta.

November 2017–April 2018: Alta-Guard observes only 288 inches of snow, the lowest seasonal snowfall on record

October 2022–April 2023: Alta Ski Area records an incredible 903 inches of snow, the most since they began collecting measurements in 1980.

2 Wasatch Microclimates

Figure 2.1. Mount Millicent and Brighton Basin in Big Cottonwood Canyon as photographed by meteorologist S. D. Green on November 26, 1931. Courtesy Special Collections Department, J. Willard Marriott Library, University of Utah.

Utah's powder reputation is built primarily on the snow climate found in the Cottonwood Canyons, a fact that was apparent to the earliest Wasatch skiers. In 1935, meteorologist and backcountry skier S. D. Green predicted that "skiers will eventually find that the Brighton Basin, or the heads of the [Cottonwood] canyons within a short radius of this winter paradise, offer the best skiing to be found in the Wasatch Mountains" (Kelner 1980, 155) (figure 2.1). Green was right. The Cottonwood Canyons are the climatological sweet spot of the Wasatch Mountains, with their own **microclimate** that produces more snow than falls in the surrounding area. On any given day, understanding the microclimates of the Wasatch Mountains can help you find the deep powder, sunshine, or spring snow that your heart craves.

LITTLE COTTONWOOD CANYON

Little Cottonwood Canyon, home to Alta, Snowbird, and extensive backcountry terrain, penetrates eastward into the Wasatch Mountains from the Salt Lake Valley (figure 2.2). It is a steep glacier-carved canyon surrounded by mountains that reach to over 11,000 feet, including the 11,498-foot American Fork Twin, the highest peak in the central Wasatch Mountains. Although Mount Timpanogos and Mount Nebo in the southern Wasatch Mountains are slightly higher, the high terrain surrounding Little Cottonwood Canyon is the most extensive in the Wasatch Mountains.

As Alta meteorologist Mike Kok once told me, "It doesn't need a reason to snow in Little Cottonwood Canyon; it needs a reason to stop." Storms don't come to Little Cottonwood to die. They come to be invigorated and rage on. After a storm ends in downtown Salt Lake City, often one can look southeastward and see it dumping in Little Cottonwood Canyon fewer than twenty miles away.

The snowfall contrasts in Little Cottonwood are among the most dramatic in the world (figure 2.3). At the canyon entrance, the average annual snowfall is about 100 inches. Above 8,500 feet, which includes most of the terrain encompassed by Alta and Snowbird, the average annual snowfall exceeds 500 inches. Above 10,500 feet, the average annual snowfall likely exceeds 600 inches,

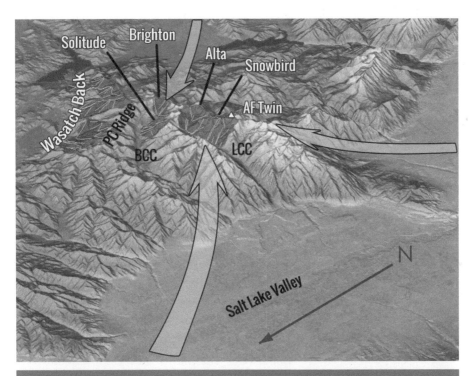

Figure 2.2. Little and Big Cottonwood Canyons (LCC and BCC, respectively) from the northwest. Exposure to flow from several directions (blue arrows) leads to precipitation enhancement in a wide variety of storms. Park City Ridgeline (PC Ridge) and American Fork Twin (AF Twin) identified with abbreviations.

There are only a few locations in the Wasatch Mountains where daily snowfall has been reliably measured for more than twenty-five years, the minimum required by most meteorologists for a long-term climatology of average annual snowfall. Snow reports from ski areas have a poor reputation and often are not collected when the ski area is closed, thus failing to account for early-season snowfall. Data collected by snow-safety professionals is typically high in quality, but in some cases is incomplete or hasn't been collected at the same site for a twenty-five-year period.

Therefore, the analysis of average annual snowfall presented in figure 2.3 was created in the following way. Since there are many more stations that report snow water equivalent (obtained from precipitation gauges) than snowfall, we began with a high-resolution analysis of average annual precipitation provided by the PRISM Climate Group at Oregon State University, which is the highest-quality precipitation analysis available in the United States. We then estimated the average annual snowfall based on how much of the annual precipitation falls as snow (which varies with elevation) and the average water content of snow in the Wasatch Mountains. The resulting analysis closely fits direct observations of average annual snowfall provided by long-term stations in and around the Wasatch Mountains.

This snowfall analysis is most uncertain at the highest elevations, where observations are not collected, and precipitation is estimated based on extrapolation from lower-elevation data. This includes the high terrain (above 9,000 feet) surrounding Little Cottonwood Canyon, although experience suggests that these estimates are reasonable. Uncertainty is also greater in areas that lack representative observations entirely, such as the area around Powder Mountain.

but nobody knows for sure because there are no brave souls taking regular measurements at such altitudes. Throughout Little Cottonwood, the average annual snowfall increases about 100 inches per 1,000 feet of elevation gain.

BIG COTTONWOOD CANYON

Big Cottonwood Canyon lies immediately to the north of Little Cottonwood Canyon (figure 2.2). The high ridge between the two canyons rises to over 11,000 feet, but the terrain to the north of Big Cottonwood Canyon, including the Park City Ridgeline, is somewhat lower, with peaks near 10,000 feet. At ski areas in upper Big Cottonwood Canyon, the average annual snowfall is just a bit lower than that found at comparable elevations in Little Cottonwood Canyon, ranging from about 400 inches at 8,000 feet (near the base of Solitude) to 500 inches at 10,500 feet (near the summit of Brighton [figure 2.3]). Average annual snowfall in the backcountry terrain on the north side of Big Cottonwood Canyon, including the Park City Ridgeline, is about 20–30 percent lower than found at comparable elevations in Little Cottonwood Canyon, reaching just over 400 inches at 10,000 feet.

THE WASATCH BACK

The Wasatch Back lies on the east side of the Wasatch Mountains and includes Park City Mountain Resort, Deer Valley, Woodward, and Mayflower ski areas, the latter scheduled to open in December 2023 (figure 2.4). At Park City Mountain Resort, the highest lifts reach to near the top of the Park City Ridgeline, where the average annual snowfall is about 400 inches (fig-

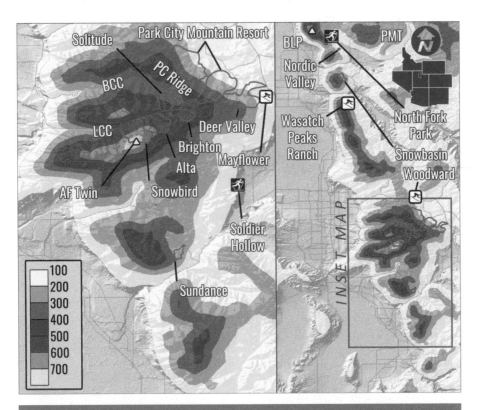

Figure 2.3. Estimated average annual snowfall in the Wasatch Mountains (in inches). Little Cottonwood Canyon (LCC), Big Cottonwood Canyon (BCC), Park City Ridgeline (PC Ridge), Ben Lomond Peak (BLP), and Powder Mountain (PMT) identified with abbreviations. Red contours indicate ski area boundaries. Data sets used in generating this analysis include those provided by the PRISM Climate Group, Oregon State University, http://prism.oregonstate.edu.

ure 2.3). At the Park City Mountain Village and Canyons Village base areas, and the Woodward ski area, however, the snowfall is less than 200 inches and about 40–50 percent lower than found at comparable elevations in Little Cottonwood Canyon. The reduced snowfall is a consequence of being on the downstream or **leeward** side of the Wasatch crest.

Deer Valley is located on a ridge that extends eastward away from the Wasatch crest (figure 2.4). Average annual snowfall is near 350 inches at the top of Empire Peak (9,570 feet), which is just a couple of miles to the lee of the Wasatch crest but decreases dramatically as one moves farther eastward and to lower elevations. Average annual snowfall at the base of the Jordanelle Gondola is probably less than 150 inches (figure 2.5). Snowfall at the base of Mayflower is comparably low.

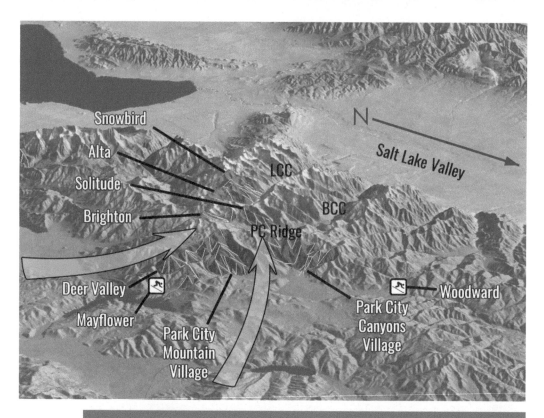

Figure 2.4. The Wasatch Back from the northeast. Although typically downstream of the Wasatch Mountains, strong precipitation enhancement can occur during low-level flow from the east and southeast (blue arrows). Park City Ridgeline (PC Ridge), Little Cottonwood Canyon (LCC), and Big Cottonwood Canyon (BCC) identified with abbreviations.

SUNDANCE AND MOUNT TIMPANOGOS

Mount Timpanogos, home of Sundance ski area, rises an incredible 7,000 vertical feet above Utah Lake (figure 2.6). Although the summit reaches higher than any peak above Little Cottonwood Canyon, the average annual snowfall is lower, attaining 500 inches only above about 11,000 feet (figure 2.3). Sundance ski area lies at lower elevations (6,100–8,250 feet) on the mountain's picturesque southeast flank and receives an average annual snowfall of about 200 inches at the base and perhaps 300 inches near the top of the lifts. The contrast in snowfall between Mount Timpanogos and Little Cottonwood Canyon is a consequence of storm climatology, terrain orientation, and terrain shape, as discussed later in this chapter.

Empire Peak
9,570 ft
1 mile east of Wasatch Crest

Jordanelle Gondola Base
6,570 ft
6 miles east of Wasatch Crest

Figure 2.5. Photos of Deer Valley's Empire Peak (top) and the Jordanelle Gondola (bottom) on March 18, 2007, illustrate the large contrasts in snowfall and snowpack that are produced by elevation and distance from the Wasatch crest.

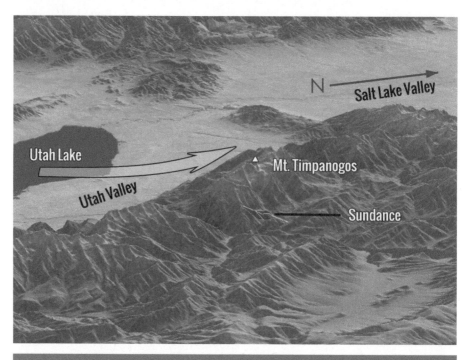

Figure 2.6. Mount Timpanogos from the southeast. Strong precipitation enhancement occurs during storms with southwesterly flow (blue arrow).

THE NORTHERN WASATCH MOUNTAINS

The Wasatch Mountains become increasingly narrow as one moves north of the Cottonwood Canyons (figures 1.3 and 2.7). The highest lifts at Snowbasin, which is located on the eastern side of the Wasatch Mountains, top out fewer than three miles from Ogden, Utah, on the western side. Although lower than Park City Mountain Resort, with elevations of 6,400 to 9,400 feet, Snowbasin is snowier, with an average annual snowfall of about 250 inches at the base and more than 400 inches on the upper mountain (figure 2.3). Just south of Snowbasin, the newly developed and private Wasatch Peaks Ranch resort receives comparable snowfall.

Farther to the north, there is a remarkable microclimate in the Ogden Valley, which lies between the northern Wasatch Mountains and the southern Bear River Range (figure 1.3 and 2.7). A small downhill ski area, Nordic Valley, operates between 5,400 and 7,000 feet on the east slopes of the Wasatch Mountains, with plans for expanding higher, and Ogden Nordic maintains a wonderful cross-country trail system at North Fork Park. Thanks to abundant low-altitude snowfall (100 to 200 inches) and colder temperatures, the snow-

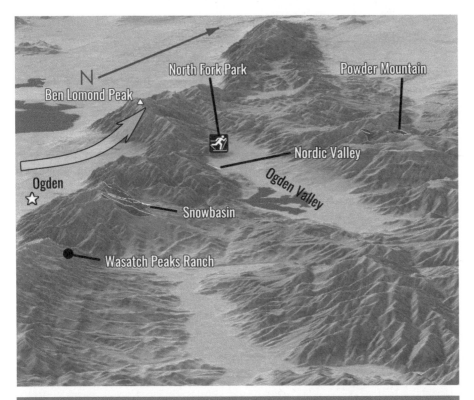

Figure 2.7. The northern Wasatch Mountains, Ogden Valley, and Powder Mountain from the southeast. Precipitation enhancement in this region occurs during storms with southerly to southwesterly flow (blue arrow), with significant spillover into the Ogden Valley.

pack in the Ogden Valley is deeper and more reliable than is found at comparable elevations in the Wasatch Mountains to the south. Meteorologically, the Ogden Valley would have been a better choice for the Olympic cross-country venue than Soldier Hollow, which lies to the lee of the formidable Mount Timpanogos and receives far less snow (figure 2.3).

Another remarkable microclimate exists just to the west of the Ogden Valley on Ben Lomond Peak, which receives heavy snowfall during storms with southwesterly flow. In fact, the average annual snowfall at 8,000 feet on Ben Lomond appears to be comparable to that at 9,700 feet at Alta. There are no lifts, but the backcountry skiing can be great (although areas above timberline are frequently windswept and prone to avalanches).

East of the Ogden Valley is Powder Mountain. Snowfall estimates for Powder Mountain are uncertain due to a lack of long-term observing stations in the area,

Snowfall averages are nice and convenient, but large variations occur from year to year, even in Utah. From 1945–1946 to 2020–2021, the average seasonal (November–April) snowfall measured by the Utah Department of Transportation near the base of Alta was 486 inches, with a minimum of 288 inches in 2017–2018 and a maximum of 745 inches in 1994–1995. Because the 2017–2018 season is the only season on record with less than 300 inches of snow, I like to tell people that a bad season at Alta is better than a good season in Colorado.

In some portions of the United States, snowfall is influenced by the seesaw of ocean temperatures in the eastern and central tropical Pacific known as the El Niño–Southern Oscillation, or ENSO. This is especially true in the Pacific Northwest where El Niño, which features higher than average ocean temperatures, is correlated with below average snowfall, and La Niña, which features lower than average ocean temperatures, is correlated with above average snowfall. There is no clear, coherent signal, however, between El Niño, La Niña, and snowfall in the Wasatch Mountains. Big snow years and low snow years have occurred in both El Niño and La Niña years. A developing El Niño or La Niña during the fall tells us little about how much snow will fall in the Wasatch Mountains during the approaching winter.

but extrapolation of the available data suggests an annual snowfall of 300 to 400 inches, although the ski area website suggests it is higher. There is always uncertainty in meteorology, so perhaps you should check it out for yourself.

UNDERSTANDING WASATCH MICROCLIMATES

Snowfall in the Wasatch Mountains reflects the storm climatology of northern Utah combined with local topographic effects. Mountains produce their own weather, and meteorologists call the snow or rain that falls when moisture-laden air is forced over a mountain **orographic precipitation** (figure 2.8). Orographic precipitation is generated on the side of a mountain range that faces the prevailing flow, known as the **windward** side. Precipitation that is carried downstream and falls out on the leeward side of the mountain range is called **spillover**. The dry area downstream is called a **precipitation shadow** and, in some cases, can also be cloud free.

During winter storms, the distribution and intensity of orographic precipitation depend on several factors, including the humidity, strength, and **stability** of the flow. The humidity and strength of the flow affect the delivery of moisture to the storm, whereas the stability determines how the flow responds to the mountains. Under stable conditions, the low-level flow near the mountains is blocked and flows along, rather than over, the mountains. This results in the development of a **blocking front** (figure 2.8), which forces the air to rise and produce precipitation upwind of the initial mountain slope. In Utah, this occurs most frequently when south-

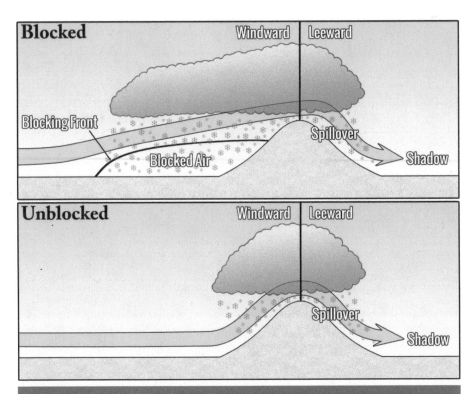

Figure 2.8. Orographic precipitation during blocked (stable) and unblocked (unstable) conditions. Adapted from Neiman et al. 2002; Cox et al. 2005.

westerly flow impinges on the northern Wasatch Mountains, generating precipitation not only for Nordic Valley, Snowbasin, and Wasatch Peaks Ranch but also the lowlands near Ogden. In contrast, during less stable conditions, the air is unblocked and rises near and over the windward slopes, resulting in precipitation primarily over the mountains.

During periods of orographic precipitation, two important phenomena can contribute to enhanced precipitation rates. The first is **orographic convection**, which is triggered when the air rising over the Wasatch Mountains becomes unstable. Although similar in many ways to thunderstorms, orographic convection in the Wasatch Mountains is usually (but not always) shallow and only infrequently produces lightning, but it still generates and enhances mountain precipitation (figure 2.9). Orographic convection produces most of the snow that falls in the Wasatch Mountains after a cold-frontal passage.

The second is known as **seeder-feeder** (figure 2.10). During seeder-feeder, air rising over the Wasatch Mountains generates a low-level "feeder" cloud that contributes to the growth of precipitation falling from "seeder" clouds aloft.

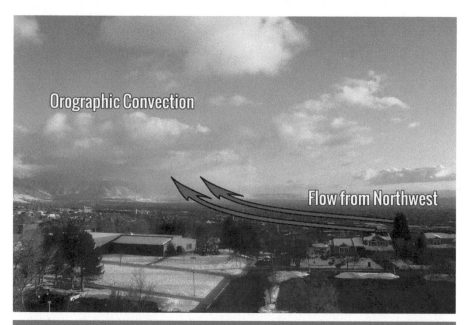

Figure 2.9. Orographic convection during northwesterly flow over the high terrain surrounding the Cottonwood Canyons on February 26, 2013. These shallow clouds produced five inches of cold smoke (less than 4 percent water content) to cap off a storm that produced a foot of snow at Alta.

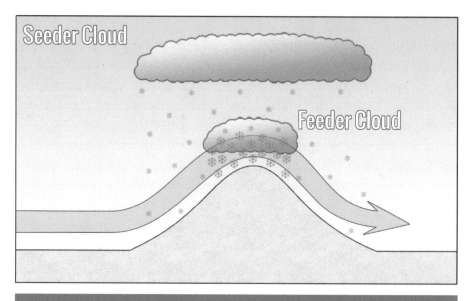

Figure 2.10. The seeder-feeder process.

Figure 2.11. Spillover and cloud and precipitation shadowing on the Park City side of the Wasatch Mountains during a shallow orographic storm on April 15, 2012 (a very poor snow year!). Photo taken looking north during westerly flow.

Seeder-feeder is most common during stable storms and can operate on very small scales, leading to an enhancement of precipitation over narrower ridges such as those surrounding Little Cottonwood Canyon.

Several factors affect the spillover of precipitation to the lee side of a mountain. On average, spillover is most pronounced downstream of narrow mountains. One reason why it snows more at Snowbasin than Park City is that there is more spillover across the lower, narrower Wasatch Mountain crest west of Snowbasin than the higher, broader topography surrounding the Cottonwood Canyons west of Park City.

Spillover is also affected by storm depth and flow strength. Shallow storms with weak flow typically produce less spillover. Some of the largest differences in snowfall between the Cottonwood Canyons and the Wasatch Back occur during periods with shallow orographic convection following the passage of a cold front. Such storms can produce prolific accumulations in the Cottonwood Canyons while there is very little snow and even some sun at the base of Park City Mountain Resort, Deer Valley, and Mayflower (figure 2.11).

A mechanism that sometimes limits the spillover of precipitation even during periods of stronger flow is the sinking of air on the leeward side of the mountains. This sinking caries snowflakes downward rapidly, which limits how far they are carried downstream (see figure 2.11). In addition, as the air sinks, it warms and dries, causing the **sublimation** of snowflakes. Sublimation is like evaporation, except the ice is converted directly to water vapor without passing through the liquid phase. Although this sounds weird, the process is the same as what happens when dry ice (frozen carbon dioxide) sublimates and "disappears" as it turns into gaseous carbon dioxide. Similarly, a snowflake can sublimate when it encounters a dry environment.

These effects contribute to the climatological decrease in snowfall from the Cottonwood Canyons on the windward side of the Wasatch Mountains to the Wasatch Back on the leeward side. Nevertheless, there are periods when the snowfall along the Wasatch Back and in the Cottonwood Canyons is similar, such as when a large-scale atmospheric feature, such as a cold front, produces most of the precipitation. In addition, during storm periods with southeasterly or easterly flow, the Wasatch Back temporarily becomes the windward side of the Wasatch Mountains and may receive more snow than the Cottonwood Canyons (figure 2.4). Flow from the south and southeast can also generate heavy snow over the west-to-east-oriented Deer Valley ridgeline, with spillover into the Park City area.

TERRAIN SHAPE AND ORIENTATION

The shape and orientation of local terrain features strongly influence snowfall throughout the Wasatch Mountains. The broad, high terrain surrounding the Cottonwoods is exposed to flow from nearly any direction, leading to precipitation enhancement in a wide variety of storms (figure 2.2). As a result, the Cottonwoods observe the greatest diversity of storms in the Wasatch Mountains. In contrast, Mount Timpanogos is very high, but also narrow. Its southwest-facing orientation favors precipitation enhancement when the flow is from the south, southwest, or west (figure 2.6), but its narrow profile does little to enhance precipitation when the flow is from the northwest. As a result, the annual snowfall on Mount Timpanogos and at Sundance ski area is lower than found at comparable elevations in the Cottonwood Canyons. For powder at Sundance, look for storms with southwesterly flow that don't have high snow levels.

The remarkable microclimate on Ben Lomond Peak also results from terrain shape and orientation. Ben Lomond Peak juts out farther to the west than the

Figure 2.12. Terrain-driven convergence and upslope flow during large-scale southwesterly flow can generate heavy snowfall on Ben Lomond and in the Ogden Valley to the east. Courtesy Tim Jansa.

rest of the northern Wasatch Mountains (figure 2.7). As a result, flow from the southwest encounters a **terrain concavity**, resulting in enhanced convergence, upslope flow, and precipitation over Ben Lomond Peak (figure 2.12). Local precipitation enhancement can also occur when the flow is blocked, moves along the northern Wasatch Mountains, and encounters Ben Lomond Peak. Some of this precipitation spills over into the Ogden Valley to the east, contributing to the remarkable low-elevation snow climate discussed early in this chapter.

For its elevation, Ben Lomond Peak is probably the snowiest location in northern Utah. I frequently take a group of students up Ben Lomond Peak in the spring to measure snow depth, and we are always impressed. Typically, we find that the snow depth at 8,000 feet is close to that observed at 9,700 feet at Alta (figure 2.13).

Figure 2.13. The author in a thirteen-foot-deep snowpack at 8,000 feet on Ben Lomond Peak on April 26, 2005.

THE GREAT SALT LAKE EFFECT

Contributing modestly to the snow climate of the Wasatch Mountains is the **Great Salt Lake effect**, which produces snowfall several times a year (figure 2.14; see also chapter 5). Given the general northwest–southeast orientation of

Figure 2.14. The author's son, Erik, takes advantage of his father's lake-effect forecast at Alta.

the lake, and the fact that cold air masses typically approach the region from the northwest, the mountains south and east of the Great Salt Lake receive the most lake-effect snow (figure 2.15). In Little Cottonwood Canyon, lake-effect periods produce about 5 percent of the average precipitation from September 16 through May 15, with lesser amounts in other regions of the Wasatch. Most people expect a larger percentage, but the marketing hype is not matched by reality. Nevertheless, the peak of the lake-effect season is in October and November, and a large lake-effect storm can help kick off the ski season. Lake-effect storms produce less precipitation along the Wasatch Back because they are shallow with little spillover.

SUBLIMATION AND MELTING OF FALLING SNOW

One additional contributor to the increase in annual snowfall with elevation in the Wasatch Mountains is the sublimation and melting of falling snow. We have

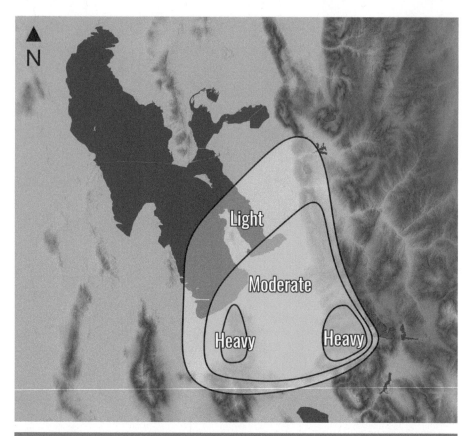

Figure 2.15. The lake-effect snowbelt of the Great Salt Lake. Background hillshade sources: Esri, USGS, FAO, NOAA.

already discussed how sublimation helps limit spillover, but there are many storm periods during which sublimation reduces snowfall in the valleys and canyons on both the windward and leeward sides of the Wasatch Mountains.

During winter, the lowlands west of the Wasatch Mountains, mountain valleys east of the Wasatch Mountains, and canyons like Big and Little Cottonwood are often filled with dry air at lower elevations. Snow that falls from aloft into this dry air sublimates, resulting in large contrasts in snowfall between the upper and lower elevations (figure 2.16). At the extreme, the snow completely sublimates before hitting the valley or canyon floor. Precipitation that falls from a cloud but doesn't reach the ground is called **virga**, and meteorologists jokingly call these situations **cloud storms**.

In addition, some of the increase in average annual snowfall with elevation in the Wasatch Mountains reflects the fact that a greater fraction of precipita-

Figure 2.16. Snowfall aloft reaching the ground at upper elevations in the Wasatch Mountains, which are obscured, but sublimating in the dry air at low levels so that little snow reaches the base of the mountains or the valley floor.

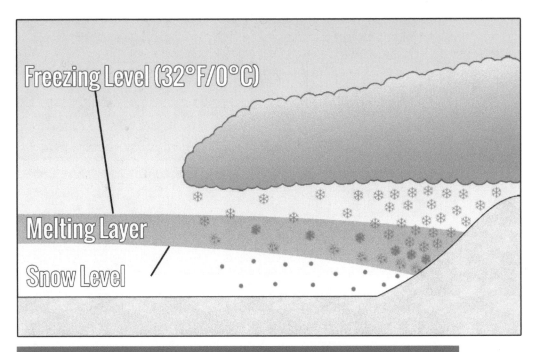

Figure 2.17. The transition of falling precipitation from snow to wet snow to slush and finally to rain beginning at the freezing level and continuing through the melting layer. The freezing and snow levels are typically lower and the melting layer deeper over the mountains than over the upstream lowlands.

Figure 2.18. John Pieper gets freshies in the foothills along Salt Lake City's east bench. Courtesy Tyler Cruickshank.

tion falls as snow at higher elevations. The **snow level** during warmer Wasatch Mountain snowstorms can sometimes approach 8,000 or even 9,000 feet, resulting in lower-elevation rain (figure 2.17). On the other hand, during cold storms, snow will fall all the way to the floor of the Salt Lake Valley, and sometimes you can ski the foothills right next to town (figure 2.18). Typically, the freezing level is lower and the melting layer deeper over the mountains, leading to a lower snow level than found over the adjacent lowlands. These effects are a consequence of several factors, including cooling produced by rising air and higher precipitation rates over the mountains.

At the base of Park City (7,000 feet), about 85 percent of the precipitation during the ski season (November–April) falls as snow. At 8,750 feet near the base of Alta, 99 percent of the precipitation during the ski season falls as snow. Thus, rain is extremely rare during the ski season at upper elevations of the Wasatch, but does fall sometimes at lower and mid elevations, resulting in less average snowfall.

3 North America

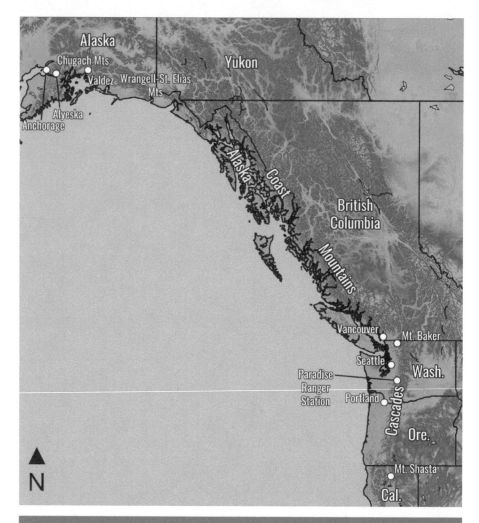

Figure 3.1. The coastal ranges of northwest North America. Background hillshade sources: Esri, USGS, FAO, NOAA.

The Wasatch Mountains have an extraordinary climate for deep-powder skiing, especially in the Cottonwood Canyons, but what about the rest of North America? Each of North America's major ski regions feature a unique snow climate and fascinating microclimates. Let's have a look.

COASTAL RANGES OF NORTHWEST NORTH AMERICA

The coastal ranges of northwest North America are among the snowiest in the world and include the Cascade Mountains of northern California, Oregon, and

Washington; the Coast Mountains of British Columbia and southeast Alaska; and the Chugach and Wrangell–St. Elias Mountains of southern Alaska (figure 3.1). Mt. Baker ski area, located near the Canadian border in the Cascade Mountains of Washington, averages about 650 inches annually and observed a world-record 1,140 inches of snow during the 1998–1999 season (July to June [figure 3.2]). A whopping 304 inches fell in February 1999, forcing the area to close for two days just to dig out the lifts. The previous world record of 1,122 inches was also set in the Cascade Mountains, at Paradise Ranger Station on Mount Rainier in 1971–1972. The single-storm world record is 189 inches, set at Mt. Shasta Ski Bowl in the Cascade Mountains of northern California from February 13–19, 1959.

Due to the strong influence of the Pacific Ocean, the coastal ranges of northwest North America feature a maritime snow climate with mild temperatures, frequent low-elevation rain, and heavy upper-elevation snowfall. Snowfall increases rapidly with elevation, a consequence of both precipitation enhancement and a greater fraction of precipitation falling as snow at higher altitudes. Many regions and ski areas in the coastal ranges average more than 400 inches of snow annually—some much more. Nevertheless, snow conditions are quite variable, and, during warm periods, rain can fall even at upper elevations.

In the Pacific Northwest, the Cascade Mountains represent a dramatic climatological divide that separates marine air to the west from continental air to the east (figure 3.3). During winter, this often results in lower temperatures and snow levels over the east slopes of the Cascades and the Columbia Basin than the west slopes of the Cascades. Major gaps in the Cascades, including the Columbia River Gorge, Snoqualmie Pass, and Stevens Pass, serve as conduits for this cold air, resulting in cold easterly flow and locally low temperature and snow levels. Ski areas benefiting from these cold easterlies include Summit at Snoqualmie and Stevens Pass. It is not uncommon for Seattle skiers to see no snow on the 4,167-foot-high Mt. Si on the warmer, western side of the Cascades, but find towering snowbanks at 3,000 feet in Snoqualmie Pass. This is a direct consequence of frequent cold easterly flow in Snoqualmie Pass, which acts as a local refrigerator, lowering snow levels during many storms and maintaining the snowpack.

Dangerous avalanche conditions can develop in these passes during frontal passages when the flow switches from cold easterly to mild westerly. The resulting increase in temperature and in some cases a rapid change from snow to rain are common triggers for avalanches. Backcountry travelers in the

Figure 3.2. A towering snowbank at Mt. Baker ski area during the world-record 1998–1999 season. Courtesy Mt. Baker ski area.

Figure 3.3. Schematic depiction of cold easterly flow in gaps and passes of the Cascade Mountains. Background hillshade sources: Esri, USGS, FAO, NOAA.

Cascades need to be attentive for such weather changes. Even if there isn't a transition to rain, the increase in temperature can lead to upside-down snow conditions and more challenging "powder" skiing conditions.

Snow consistency and quality does improve somewhat as one moves northward. Although vulnerable to the occasional rainstorm, the Chugach Mountains of southern Alaska observe drier snow on average than the coastal ranges farther to the south.

The only major ski area in Alaska's Chugach Mountains is Alyeska. Alyeska has a base elevation of only 250 feet, with lifts to 2,750 feet, although those willing to earn their turns can climb to the Mount Alyeska summit at 3,939 feet. The average annual snowfall at Alyeska increases dramatically from about 210 inches at 250 feet near the base to just over 500 inches at 1,400 feet. Snowfall is even greater on the upper mountain. This remarkable contrast reflects not only the increase of precipitation with altitude but also a greater fraction of precipitation falling as snow instead of rain.

Similarly, prolific snow falls in the heli-skiing region around Valdez, Alaska. The world-record two-day snowfall of 120.6 inches was set at Thompson Pass just east of Valdez on December 29–30, 1955. During that storm, the three-, four-, and five-day accumulations were 147, 163, and 175.4 inches, respectively.

Although the snow is drier on average than in the coastal ranges to the south, the challenge in the Chugach is the short day length during the high-latitude winter and the dramatic, rapid variability in weather and snow conditions arising from the battle between maritime air masses from over the Pacific Ocean and continental air masses from Alaska's interior. You could have the greatest skiing of your life one day and your worst the next. Bring your snorkel but be prepared for anything.

CALIFORNIA'S SIERRA NEVADA

The Sierra Nevada occupy most of eastern California, with Lake Tahoe roughly marking the transition between the lower northern Sierra and the much higher southern Sierra (figure 3.4). The Sierra Nevada have a predominantly maritime snow climate, but one that is punctuated by fewer clouds and longer dry periods than found in the Cascade Mountains to the north. This "feast or famine" snow climate is punctuated by large multiday blizzards, extended dry spells, and large month-to-month and year-to-year variations in snowfall. Tamarack, California, which is located at about 7,000 feet just south of Lake Tahoe, observed 313 inches of snow during March 1907, still the US record for snowfall in a calendar month.[1] Seasonal snowfall measured near 9,000 feet at Mammoth Mountain has varied from as little as 94 inches in 1976–1977 to as much as 661 inches in 2010–2011.

The snowiest ski areas are located along the Sierra crest near Lake Tahoe and include Palisades Tahoe, Kirkwood, Alpine Meadows, and Sugar Bowl, all of which average over 400 inches annually at upper elevations. Snowfall is also abundant at high elevations in the southern Sierra. Much of the southern Sierra is difficult to access in the winter, but the region is famous for outstanding backcountry corn skiing in the spring.

Mammoth Mountain in the southern Sierra is home to one of the more remarkable snow microclimates in the western United States. It is fortuitously located at the head of the San Joaquin River Valley, which forms a deep channel in the otherwise formidable topography of the southern Sierra (figure 3.4). Moist Pacific air masses penetrate up this channel and generate substantial precipitation, mostly in the form of snow, when they encounter Mammoth Mountain.

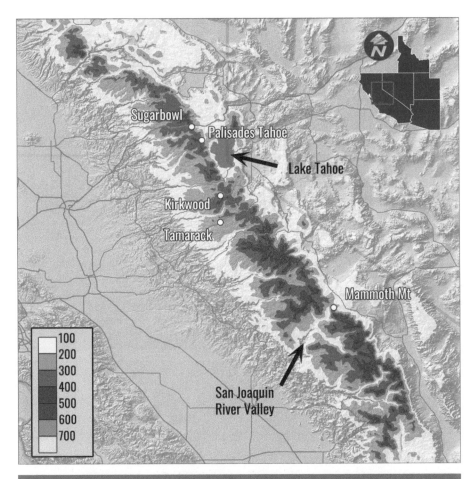

Figure 3.4. Estimated average annual snowfall of the Sierra Nevada (in inches). Data sets used in generating this analysis include those provided by the PRISM Climate Group, Oregon State University, http://prism.oregonstate.edu.

Long-term records collected by the US Forest Service and Mammoth Mountain ski patrol suggest an average annual snowfall of 353 inches at 9,000 feet. Average annual snowfall near the summit (11,000 feet) may exceed 500 inches.

A large fraction of the precipitation that falls in the Sierra Nevada is produced by **atmospheric rivers**, narrow filaments of moisture that impinge on California from the tropics and subtropics (figure 3.5). As a result, the Sierra Nevada are famous for large multiday snowstorms with high water contents. For example, an atmospheric river produced 91.5 inches of snow with an average water content of 12 percent at Mammoth Mountain from December 18–20, 2010.

Curiously, the snow is "heavy" in another way at Mammoth Mountain. Snow is comprised of water molecules, which include two hydrogen atoms

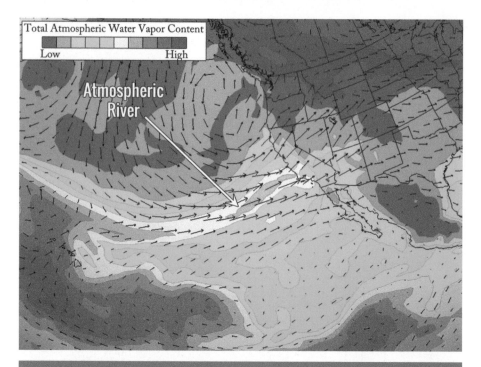

Figure 3.5. A narrow filament of tropical and subtropical moisture, known as an atmospheric river, streams toward the Sierra Nevada at 4:00 p.m. PST on December 19, 2010. This atmospheric river produced more than ninety inches of snow at Mammoth Mountain in seventy-two hours.

and one oxygen atom. Hence the chemical formula H_2O. Most hydrogen atoms contain a positively charged proton and a negatively charged electron, but a small fraction of them also contain a neutron. Hydrogen atoms with a neutron are known as **deuterium**, or heavy hydrogen. Deuterium is harmless, naturally occurring, and represents about one in every 6,240 hydrogen atoms on Earth.

Water molecules containing deuterium are slightly heavier than those containing regular hydrogen and condense more readily into water or ice. As storms move inland from the Pacific Ocean and over the Sierra Nevada, there is a tendency for these heavier water molecules to rain out or snow out faster than regular water molecules. As a result, the amount of deuterium in Sierra snow gradually decreases as you move inland, and snowfall on the east side typically contains less deuterium than the west side.

The area around Mammoth Mountain, however, is a special case because moist Pacific air masses produce less precipitation as they penetrate up the San Joaquin River Valley. As a result, the snow in Mammoth is relatively enriched in

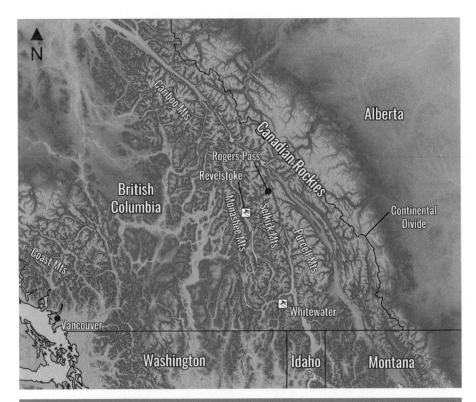

Figure 3.6. The Columbia Mountains (i.e., the Cariboo, Monashee, Selkirk, and Purcell Mountains) and Canadian Rockies. Background hillshade sources: Esri, USGS, FAO, NOAA.

deuterium compared to the surrounding eastern Sierra. This has a negligible impact on the density of snow at Mammoth Mountain, but it does provide an atomic "fingerprint" that allows scientists to better understand the meteorology and hydrology of the region.

COLUMBIA MOUNTAINS AND CANADIAN ROCKIES

As one moves inland, the wintertime climate of western North America generally becomes drier, but snow quality and consistency improve. The best powder skiing is found in areas where there is both snow quantity and quality. The Wasatch Mountains of northern Utah are one such area. Another is the Columbia Mountains of British Columbia and far northern Washington, Idaho, and Montana, which include the Cariboo, Monashee, Selkirk, and Purcell Mountains (figure 3.6). This mountainous region is one of the few that Utah powder snobs flock to for a lift-served, heli, or backcountry ski vacation.

The Columbia Mountains are massive in scope and scale. The heli-ski area serviced by Canadian Mountain Holidays in the Columbia Mountains equals that of the entire Swiss Alps! There are snowbelts, banana belts (relatively warm, sunny areas), and a host of microclimates too numerous to mention. In general, average annual snowfall in most upper-elevation areas (above 5,000 feet) approaches or exceeds 350 inches, but it can be much greater in favored areas. Long-term records from Mount Fidelity near Rogers Pass in the Selkirk Mountains indicate an average annual snowfall of 564 inches at 6,150 feet. For lift-served skiing, snowier areas include Whitewater and upper-elevation terrain at Revelstoke.

The Columbia Mountains are influenced by both maritime and continental air masses. At times, rain and **rime**, the latter a form of icing produced by clouds or drizzle freezing on contact with ground-based objects and snow, affect ski conditions. Extended periods of cold, dry weather can also occur.

Owing to the loss of moisture to upstream mountain ranges, average annual snowfall generally decreases at a given elevation as one moves east (inland). Snowfall in the Canadian Rockies is substantially lower than in the Columbia Mountains, especially in areas east of the Continental Divide.

TETON RANGE

The Teton Range of Wyoming and Idaho (figure 3.7) has a transitional snow climate like that found in the Wasatch Mountains, resulting in abundant high-quality snow. Snowfall is greatest near and west (windward) of the Teton crest. Grand Targhee on the windward western side of the Tetons averages about 475 inches seasonally (November–April) at the base, placing it just behind Alta. Snowfall at comparable elevations at Jackson Hole Mountain Resort on the leeward eastern side of the Tetons is lower, exceeding 400 inches above about 9,400 feet and declining sharply to less than 200 inches at the base. This contrast in snowfall combined with an overall southeastern exposure can yield dramatic differences in snow conditions between the upper and lower mountain.

COLORADO ROCKIES

Many Colorado ski areas receive drier snow than Utah, but even near the top of their lift-served terrain, their average annual snowfall is lower than at ski areas in the Cottonwood Canyons (figure 3.8). Mean annual snowfall at 9,200 feet at Steamboat Springs is about 380 inches, possibly exceeding 400 inches near the

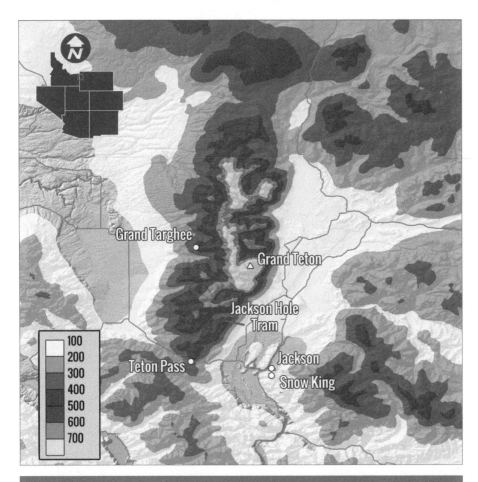

Figure 3.7. Estimated average annual snowfall (in inches) of the Teton Range. Data sets used in generating this analysis include those provided by the PRISM Climate Group, Oregon State University, http://prism.oregonstate.edu.

summit, but is only about 200 inches at the base. Beaver Creek, Winter Park, and Vail average about 350 inches near their summits but, as with Steamboat, snowfall is more meager at their bases. Summit County ski areas like Copper Mountain, Breckenridge, and especially Keystone are drier. Thanks to altitude and proximity to the Continental Divide, Arapahoe Basin and Loveland Pass average over 300 inches at their bases, which are above 10,000 feet. With less snowfall and a climatology that favors smaller storms, the frequency of deep-powder days is considerably lower at these Colorado ski areas than in the Cottonwood Canyons.

Analyses of estimated average annual snowfall for Colorado suggest that snowfall is greatest in three regions (figure 3.8). The first is in the San Juan

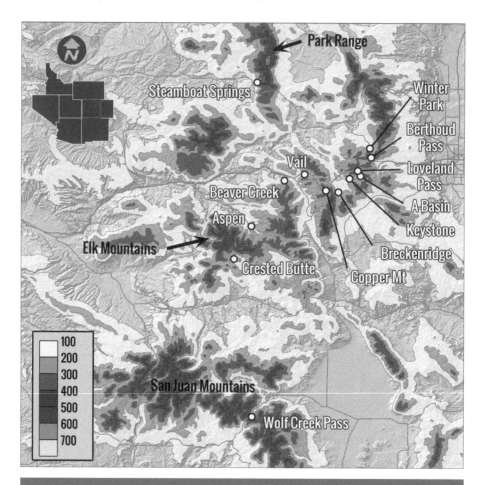

Figure 3.8. Estimated average annual snowfall (in inches) of the Colorado Rockies. Data sets used in generating this analysis include those provided by the PRISM Climate Group, Oregon State University, http://prism.oregonstate.edu.

Mountains near Wolf Creek Pass and Wolf Creek Pass ski area. Long-term observations from this region indicate an average annual snowfall of 435 inches at 10,650 feet. Most of the snow at Wolf Creek Pass falls during warmer, south-westerly flow storms and has a higher average water content (10.3 percent) than found elsewhere in Colorado (typically about 7–8 percent). The second is in the rugged high-elevation backcountry of the Elk Mountains near Aspen and Crested Butte. Thanks to elevation and exposure, this area receives more snow than the Aspen area and Crested Butte ski areas. The third is in the remote Park Range north of Steamboat Springs. Snowpack observations collected along this high, north-south-oriented barrier suggest that snowfall may average 600 inches above 10,000 feet.

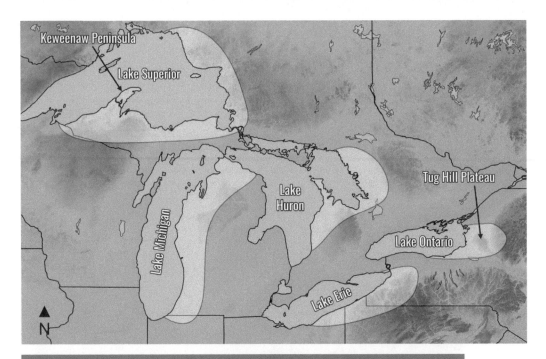

Figure 3.9. Lake-effect snowbelts of the Great Lakes. Background hillshade sources: Esri, USGS, FAO, NOAA.

EASTERN UNITED STATES

Ski conditions in the eastern United States are strongly impacted by winter thaws, rain, and storms that produce sleet and freezing rain. Skiing here is not for the faint of heart! Nevertheless, when conditions are right, powder can be found, especially in the snowbelts downstream of the Great Lakes (figure 3.9). Portions of the Tug Hill Plateau of New York and the Keweenaw Peninsula of Michigan average over 200 inches annually, much of it falling from November through February when the lakes are ice free and relatively warm. On the Keweenaw Peninsula, Mt. Bohemia combines the best of what is possible from the regional snow climate and topography. It sits near the south shore of Lake Superior, claims an average snowfall of 273 inches, and offers up the most vertical and powder in the Midwest United States. They even offer cat skiing on nearby Voodoo Mountain.

While growing up in upstate New York, my first introduction to deep-powder skiing was at Snow Ridge, a ski area in the snowbelt east of Lake Ontario (figure 3.10). Snow Ridge may be small, but it sits on the eastern edge of the Tug

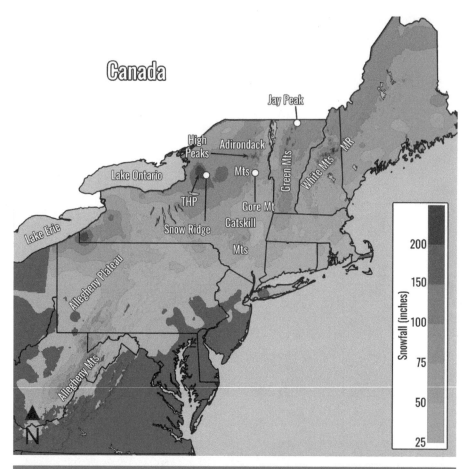

Figure 3.10. Average annual snowfall in the northeast United States. Tug Hill Plateau (THP) and Mahoosuc Range (MR) identified with abbreviations. Background hillshade sources: Esri, USGS, FAO, NOAA. Snowfall data source: NOAA/National Weather Service Forecast Office in Burlington, VT.

Hill Plateau where snowfall produced by Lake Ontario maximizes. The Tug Hill Plateau unfortunately contains few skiable slopes except on its east side where an escarpment drops abruptly into the Black River Valley. Snow Ridge is located where the drop of this escarpment is the steepest and the longest. The vertical drop is only 500 feet, but with no real mountains nearby, Snow Ridge provides the best option for deep-powder skiing. Call it the Alta of upstate New York. Ski areas in far western New York also benefit from relatively abundant lake-effect snowfall, especially where the Allegheny Plateau rises above Lake Erie. Farther south, the Allegheny Mountains of West Virginia represent the snowiest region of the mid-Atlantic states and can do well when there is north-westerly upslope flow of lake-modified air.

Elsewhere in the northeast United States, snowfall maximizes in the Adirondack High Peaks, along the spine of the Green Mountains of Vermont, and in the White Mountains and Mahoosuc Range of New Hampshire and Maine (figure 3.10). Jay Peak in northern Vermont is well positioned geograph-ically and topographically to benefit from a diversity of storms from coastal lows known as nor'easters to unstable northwesterly flows and enjoys a cult-like status among eastern hardcore powder skiers.

One natural snowfall enigma is Gore Mountain in the eastern Adirondack Mountains. Gore is the largest ski area in New York by acreage and miles of trails, but it is too far east to strongly benefit from lake-effect storms and too far inland to strongly benefit from nor'easters. While it can get something from each of these storm types, it's not enough to bring seasonal snow average up to that of other large northeast resorts, especially those in favored areas of Vermont, New Hampshire, and Maine.

NOTE

1. There is some debate about the legitimacy of this record among meteorologists. There is no debate, however, that the Sierra Nevada experience some of the largest multiday snowstorms in the United States.

4 The World

It's a Herculean task to summarize the snow climates of the world. Given the right conditions, you can find deep days from the European Alps to Japan's Gosetsu Chitai (heavy snow region). So many microclimates, so little time! Here's a CliffsNotes-like summary of snow climates beyond North America.

EUROPEAN ALPS

The European Alps are the birthplace of Alpine skiing and home to the world's biggest ski areas. Unlike the mountain ranges of western North America, which run primarily from north to south, the Alps run from east to west with a hook on their western end (figure 4.1).[1] Storms impinge on the Alps from the Atlantic, Mediterranean, northern Europe, and eastern Europe, so there are less distinct windward and leeward effects on their average annual precipitation climatology compared to the predominantly north-south-oriented ranges of western North America. Instead, the windward and leeward sides of the Alps vary from storm to storm. For example, southerly flow can produce heavy precipitation in the southern Alps one day, while northwesterly flow generates heavy precipitation in the northern Alps the next. Different regions of the Alps can have dramatically different ski conditions, depending on the characteristics of recent storms.

The Alps are also so high and wide that many storms dump their loads on their periphery, sometimes called the **Alpine Rim**, with less precipitation over the interior. As a result, precipitation is greatest along the northern Alpine Rim from eastern France to central Austria and along the southern Alpine Rim from northwest Italy to western Slovenia (figure 4.1). The greatest wintertime snowfall is found at high elevations in these regions (at low elevations, a greater fraction of precipitation falls as rain, especially along the southern rim). In contrast, despite their high elevations, the Valais (a.k.a. Pennine) Alps of Switzerland and Italy and the interior eastern Alps from Trento to Innsbruck and St. Moritz to Lienz, including the Ötztal Alps and the interior Dolomites, are drier and observe less wintertime snowfall at a given elevation.

In relatively wet areas of the eastern Alps, snowfall at upper elevations (above 3,000 meters or 10,000 feet) can be impressive. At 3,106 meters (10,190 feet) along the Alpine divide, the median[2] annual snowfall at the Sonnblick Observatory is 828 inches! However, the snow accumulation season at this altitude is very long, and snow can fall any time of year, so this includes snowfall even in the summer months. Still, from November to April the median is 575 inches.

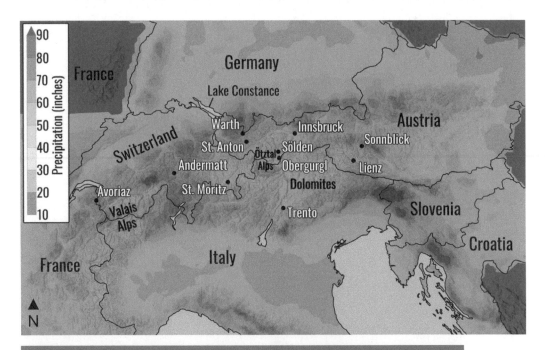

Figure 4.1. Average annual precipitation of the European Alps (in inches). Precipitation data from Isotta et al. (2014, DOI 10.18751/Climate/Griddata/APGD/1.0). Background hillshade sources: Esri, USGS, FAO, NOAA.

One of the best bets for lift-served deep-powder skiing in the Alps is in western Austria where one of the largest interconnected ski areas in the world extends through the Arlberg region from Warth to St. Anton. This is a relatively snowy region near the northern Alpine Rim where storms are forced to rise over the Arlberg massif and every now and then get a boost from lake effect generated by Lake Constance. Warth advertises a mean annual snowfall of just over 400 inches and is considered one of the snowiest ski areas in the Alps. Snowfall decreases as one moves southeastward to St. Anton but is still plentiful by Alpine standards, although a good deal of the lift-served terrain above St. Anton has southern exposure.

Farther southeast in the Ötztal Alps it becomes drier despite the terrain getting higher. This is one of the paradoxes of the Alps. Snowfall is most plentiful on the northern Alpine rim where elevations are lower and snow conditions are more variable, but over the interior, resorts like Sölden and Obergurgl reach higher elevations but receive less snowfall. On the other hand, the colder, higher elevation ski terrain at these resorts enable more consistent snow condi-

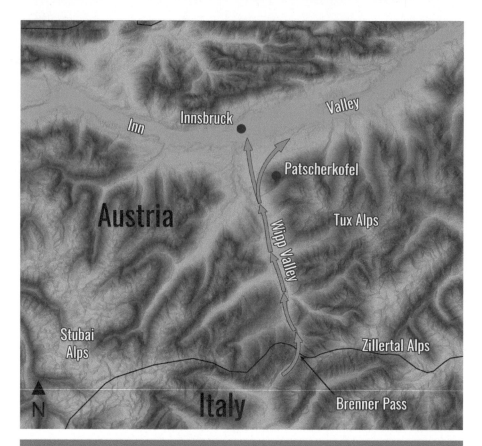

Figure 4.2. Schematic depiction of the föhn in the Wipp Valley south of Innsbruck. Background hillshade sources: Esri, USGS, FAO, NOAA.

tions. A similar snow climate transition occurs from the southern Alpine Rim to the interior Dolomites.

In the central Alps, the region surrounding Andermatt, Switzerland, has one of the more reliable snow climates because it is exposed to storms from both sides of the Alps. Annual snowfall at nearby St. Gotthard Pass is 400 inches at 6,800 feet and almost 600 inches at 7,500 feet near the top of the Nätschen ski area (known as Gütsch). While these are impressive totals, the snow accumulation season at these high elevation locations is long. The November to April snowfall at these sites is 340 and 470 inches, respectively. In the western Alps, Avoriaz on the French side of the massive Portes du Soleil resort is one of the snowier ski areas, although it also sees greater weather variability than drier, upper-elevation regions closer to the Alpine divide along the French–Italian border.

When skiing in the Alps, beware of the **föhn**. In English, it is usually spelled *foehn*, but I prefer the German föhn because, for all intents and purposes, it is a dirty four-letter word. The föhn is a warm downslope wind that occurs on the north or south side of the Alps and can produce warm temperatures, rapid snowmelt, and high winds, making the skiing less enjoyable.

One of the more vulnerable ski areas to föhn is Patscherkofel south of Innsbruck (figure 4.2). Franz Klammer became a skiing immortal when he won the downhill gold medal on the slopes of Patscherkofel during the 1976 Winter Olympic Games. I suspect there was no föhn that day or they would not have been able to hold the race. Patscherkofel sits adjacent to the Wipp Valley where it joins the Inn Valley at Innsbruck. During föhn, strong winds come down the Wipp Valley from Brenner Pass, a low gap between the Stubai Alps to the west and the Tux and Zillertal Alps to the east. On föhn days, Patsherkofel is highly exposed to strong winds, and lift disruptions or closures are common. On low wind days, however, proximity to Innsbruck and easy access via public transit make Patscherkofel a quick option for a few laps on the Olympia Downhill or a ski tour to a mountain hut to enjoy the views and Tyrolean gastronomy.

SOUTHERN ALPS OF NEW ZEALAND

New Zealand is home to the Southern Alps, which rise 6,000 to 10,000 feet out of the Tasman Sea and squeeze moisture out of storms embedded in the prevailing westerly flow throughout the year (figure 4.3). Much of this precipitation falls as rain near sea level, but heavy snow falls at high elevations along the Main Divide, which forms the spine of the Southern Alps. Most of New Zealand's major commercial ski areas are located, however, at modest elevations on the drier eastern side of the Southern Alps (e.g., Mt. Hutt, Treble Cone, Coronet Peak, The Remarkables). Snowfall at these areas is meager (generally 200 inches or less), and rain sometimes falls during winter. On the North Island, New Zealand's largest ski area, Whakapapa, sits with its sister ski area, Turoa, on the slopes of Mount Ruapehu, a spectacular volcano. Although Turoa reported a fifteen-foot base during the Southern Hemisphere spring of 2008, a record for a New Zealand ski area, the average annual snowfall is reportedly less than 200 inches.

Long-term meteorological records have never been collected at high elevations along the Main Divide, but the snowfall is prolific. There are more than 3,000 glaciers in New Zealand, including the Tasman Glacier, which flows more than ten miles along the eastern flank of Aoraki/Mount Cook, the highest

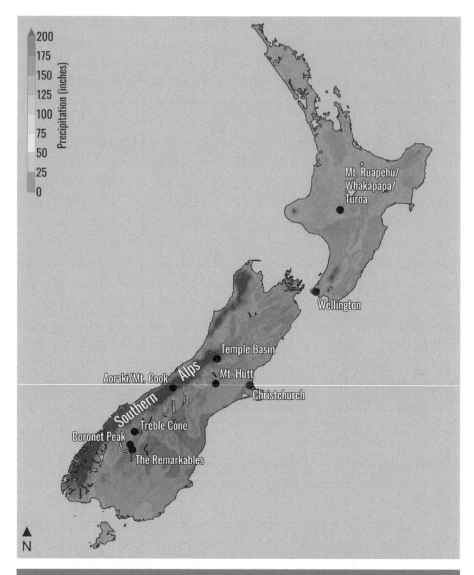

Figure 4.3. Average annual precipitation of New Zealand (in inches). Precipitation data from NIWA and Environmental Reporting, Ministry for the Environment and Statistics (New Zealand). Background hillshade sources: Esri, USGS, FAO, NOAA.

peak in the Southern Alps at 12,316 feet (figure 4.4). If you are looking for snow in New Zealand, be near the Main Divide. An option is Temple Basin, a club ski area in Arthur's Pass National Park that operates three rope tows between 4,350 and 5,750 feet and is a favorite stop for adventurous skiers and snowboarders. Be warned, though. The ski area requires a forty-five-minute walk just to get

Figure 4.4. Aerial view of the upper half of the Tasman Glacier in New Zealand's Aoraki/Mount Cook National Park. Source: Avenue, Wikipedia Commons, CC BY-SA 3.0.

the tows, which they aptly describe as "not a bad thing as the views are worth savoring." Backcountry and heli-skiing are also possible, and Mount Cook Ski Planes and Helicopters provides access to the Tasman Glacier. Nevertheless, while the snowfall is prolific, the mild maritime climate produces snow that is wetter and heavier than that found over interior continental areas such as Utah.

THE ANDES

The Andes form an incredible mountain barrier that runs the entire length of South America from the tropics to the southern tip (figure 4.5). A dramatic climate transition occurs along the Andes near Santiago, Chile, which roughly separates the "dry" Andes to the north from the "wet" Andes to the south. Despite reaching incredible altitudes, with few passes lower than 13,000 feet and several peaks over 19,000 feet, precipitation is scant over the dry Andes, which

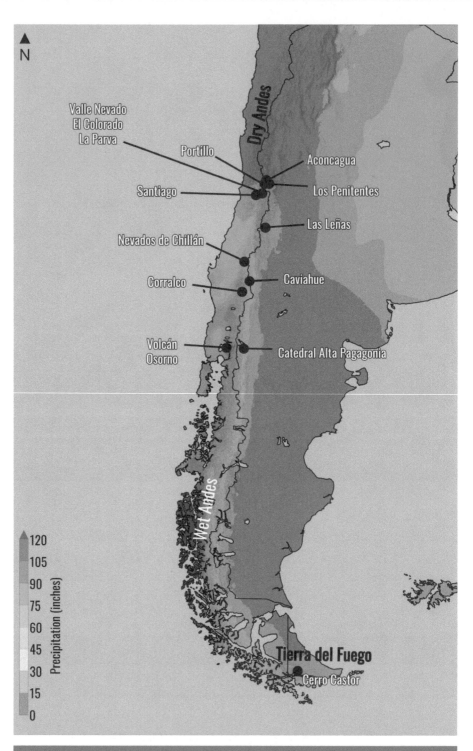

Figure 4.5. Average annual precipitation of the Andes (in inches). Precipitation data from worldclim.org. Background hillshade sources: Esri, USGS, FAO, NOAA.

sit almost continually under high pressure. Lack of oxygen, limited snow, and the tropical sun make this a less than desirable powder climate. In contrast, the prevailing westerly flow found in the mid-latitudes farther to the south generates abundant precipitation on the windward western slopes of the wet Andes, with drier conditions to the lee, much like that found in New Zealand.

Some of the largest South American ski areas, including Portillo, Valle Nevado, El Colorado, and La Parva in Chile and Los Penitentes and Las Leñas in Argentina, are located near the transition zone between the dry and wet Andes. The climate here is similar to California, with wet winters and dry summers, and the mountain climate resembles that of the southern Sierra Nevada. With elevations between 8,000 and 12,000 feet, these areas report annual snowfalls of 240–325 inches. Snowfall at Las Leñas on the lee side of the Andes varies dramatically from base to summit. Although this region has the best powder reputation in the Southern Hemisphere, it still falls short of the Cottonwood Canyons in terms of quantity and consistency.

As one moves southward into the wet Andes, the climate becomes more like the Cascades of the Pacific Northwest and, near Tierra del Fuego and South America's southern tip, where Cerro Castor is the world's southernmost ski area, maritime southeast Alaska. The height of the Andean crest also decreases from about 16,000 feet near Santiago to about 6,000 feet near Tierra del Fuego. Reliable meteorological observations from the upper elevations of the wet Andes are essentially nonexistent, although ski areas and backcountry on the Chilean side of the Andes are certainly characterized by a maritime snow climate resembling that in the Cascades. The wintertime climate on the Argentinean side is colder and drier, perhaps enabling a transitional snow climate, but snowfall is also more limited as the Andean rain shadow is one of the world's most dramatic. For example, the upper elevations of Catedral Alta Patagonia reportedly receive an average of 200 to 250 inches each winter.

JAPAN'S GOSETSU CHITAI

One of the world's great snow climates exists in Japan's **Gosetsu Chitai**, a heavy snow area on the Sea of Japan side of Honshu and Hokkaido Islands where incredible **sea-effect snowstorms** are produced when cold air from Siberia spills across the Sea of Japan during the winter monsoon (figure 4.6). Honshu Island's Hokuriku region, which lies downstream of the widest portion of the Sea of Japan, is one of the snowiest densely populated regions of the world. The city of Jōetsu, located on the Sea of Japan coast with a population of more

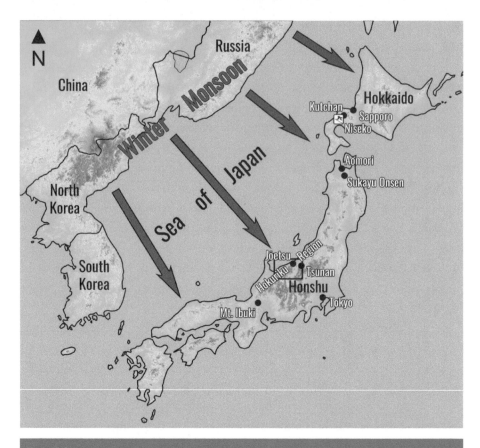

Figure 4.6. Geography and major mountain ranges of Japan. Rectangle identifies location of figure 4.7. Background hillshade sources: Esri, USGS, FAO, NOAA.

than 200,000, averages about 250 inches of snow annually, recorded an incredible 362 inches during January 1945, and once observed 59 inches in twenty-four hours, more than the twenty-four-hour record for Alta.

Southeast of Jōetsu, the Japanese Meteorological Agency observing site in Tsunan observes an average annual snowfall of 531 inches at an altitude of only 1,483 feet (figure 4.7). The Myoko-Kogen ski resorts, which provide lift-served skiing between 2,400 and 6,000 feet, benefit from proximity to the Sea of Japan and likely average about 500 inches of snow annually. Substantial snows also fall in the Hida Mountains, sometimes called the Northern or Japanese Alps, which are crossed by the Tateyama–Kurobe Alpine Route and its famous Mount Tateyama snow corridor (figure 4.8). Happo-One ski area in the Hakuba Valley east of the northern Hida Mountains hosted the Olympic downhill and Super G races during the 1998 Nagano Winter Games and is one of the largest ski

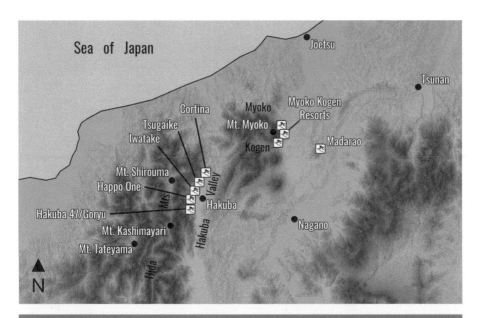

Figure 4.7. Geography of the Hakuba Valley region. Background hillshade sources: Esri, USGS, FAO, NOAA.

Figure 4.8. The Mount Tateyama snow corridor along the Tateyama–Kurobe Alpine Route through the Japanese Alps. Source: Uryah, Wikipedia Commons, CC BY-SA 3.0.

Figure 4.9. Peter Veals getting the goods in the Hida Mountain backcountry, Japan.

resorts in Japan. Farther southwest, a world-record seasonal snow depth of 466 inches was measured on Mount Ibuki in 1927 (see figure 4.6 for location).

The inland penetration of sea-effect storms varies, especially in the Hokuriku where substantial topography exists. **Satoyuki** storms produce heavier lowland (coastal) snowfall, and **Yamayuki** storms produce heavier mountain snowfall. Monitoring the inland penetration of sea-effect storms can be important for finding powder. For example, the Hida Mountains and the Hakuba Valley to their east extend southward and inland from the Sea of Japan (figure 4.7). The

Hida Mountains rise about 7,000 feet above the valley, with Mount Shirouma reaching 9,619 feet. The highest resorts reach about 6,000 feet and, while Japan is sometimes derided as "deep not steep," that's not the case in the rugged Hida Mountain backcountry, especially at upper elevations (figure 4.9).

Climatological snowfall in the Hida Mountains and adjoining Hakuba Valley is plentiful but decreases with inland extent. At the main Hakuba Valley resorts, average annual snowfall is greatest at Cortina, which is closest to the Sea of Japan and least plentiful at the interconnected Hakuba 47/Goryu resorts farther inland. Although you might assume that Cortina is always the best option, that's not always the case. Cortina may be the best option during smaller storms that feature weaker flow, which typically leads to less inland storm penetration and a sharp inland decline in sea-effect snowfall. On the other hand, in larger storms, the inland resorts may be the better option if one is dealing with "too much of a good thing" at Cortina or if you just want to ski bigger resorts. A challenge during some storms with greater inland penetration, however, can be strong winds, which can result in lift closures.

To ski tour at upper elevations it is possible to buy a single-ride lift pass and join the skin-track conga line to the Hida Mountain backcountry above some of the resorts. Keep in mind, however, that bluebird conditions with low avalanche danger are the exception and not the rule during peak powder season due to the copious snowfall and exposure of the upper elevations to the full blunt force of flow from the Sea of Japan. Tree skiing at moderate to low elevations is often the best option, although even in low-angle terrain, deep-snow immersion is a suffocation threat, especially in tree wells, brushy terrain, and terrain traps.

The Achilles heel for powder skiing in the Hokuriku and surrounding region is the low latitude and altitude. At 37°N, the Hokuriku region lies at the same latitude as San Francisco. (Imagine if San Francisco got 250 inches of snow a year and the Bay Area mountains 500 inches!) The average freezing level in January is about 1,500 feet and, given that many of the ski areas top out below 6,000 feet, the snow here can be heavy.

Farther north in Japan, however, it's colder, and you'll find drier snow and the strongest contenders to Utah's claim to the Greatest Snow on Earth. On northern Honshu, the sea-level city of Aomori averages 305 inches of snow a year (figure 4.6). South of Aomori, Sukayu Onsen in the Hakkoda Mountains is the snowiest inhabited location on Earth with a mean annual snowfall of 694 inches! Hokkaido Island is home to the Niseko United ski resorts, famous for deep powder. A Japanese Meteorological Agency climate station in Kutchan,

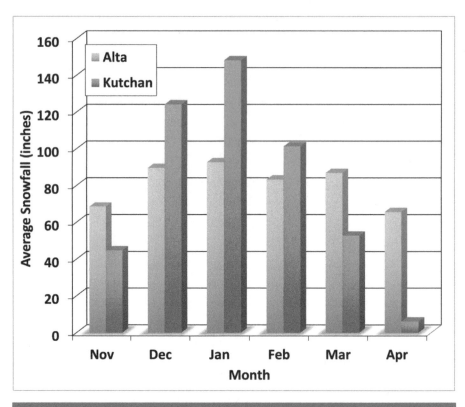

Figure 4.10. Average monthly snowfall at Kutchan and Alta. Sources: Utah Department of Transportation, Utah Avalanche Center, and Japanese Meteorological Agency.

located at an altitude of 590 feet near the Niseko resorts observes an average annual snowfall of 480 inches. As is the case throughout Japan's Gosetsu Chitai, most of this snow falls from December through February, with an average peak in January of almost 150 inches, substantially more than Alta (figure 4.10). This makes January in the mountains of northern Honshu and Hokkaido Islands near the Sea of Japan the surest climatological bet for deep-powder skiing in the world. Based on past records at Kutchan, for example, there's a 90 percent chance of at least 100 inches of snow in January.

THE REMOTE AND UNDISCOVERED

There are certainly great snow microclimates not described in this chapter, some of which remain undiscovered or largely unknown even today. Meteorological records are sparse in many regions. If you find your Shangri-la, let me know—but otherwise keep it quiet.

NOTES

1. Because of the unavailability of digital snowfall analyses, maps for the European Alps (figure 4.1), New Zealand (figure 4.3), and South America (figure 4.5) present average annual precipitation, the sum of rainfall and the snow water equivalent of snow and other frozen precipitation events.

2. The median is the value that separates the lower and upper halves of all values. It is thus in the middle of all prior observations.

5 | Flaky Science

Figure 5.1. Stellar dendrites (left) come in many forms, but all have six treelike arms radiating away from the center. Graupel (right) can be lump (as pictured), cone, or hexagonal shaped.

A million billion snowflakes fall every second for your skiing pleasure. The shape, size, and composition of these snowflakes determine if you are skiing hero snow, crud, or concrete. If you want blower pow, you're looking for **stellar dendrites**, which have six treelike arms, lots of cavities and pores, and a low water content (figure 5.1). On the other hand, maybe you are partial to **graupel**, which looks and feels like a Styrofoam ball and has a high water content but behaves like ball bearings and provides a unique, creamy ski experience. There are ten snowflake types in the simplest classification system used by meteorologists and eighty in one of the more complex. How does Mother Nature manufacture so many varieties?

MOTHER NATURE'S FIVE-STEP PLAN FOR A SNOWSTORM

Many people believe that snowflakes form when raindrops freeze, but this process produces pellets of ice known as **sleet**. Instead, Mother Nature employs a five-step plan to squeeze water vapor out of the atmosphere and dump it out as snowflakes on your favorite ski hill. These five steps are condensation, glaciation, vapor deposition, riming, and aggregation.

Condensation

The first step to produce a snow-storm is to create a cloud. Clouds are produced by **condensation**, which occurs when there is a net flow of water from vapor to liquid. For example, dew forms when atmospheric water vapor condenses into water droplets on a cold surface, such as grass or your windshield. The temperature at which this occurs is known as the **dewpoint**. When the temperature and the dewpoint are the same, the relative humidity is 100 percent, and the atmosphere is at **saturation**. The atmosphere is **supersaturated** if the relative humidity is greater than 100 percent.

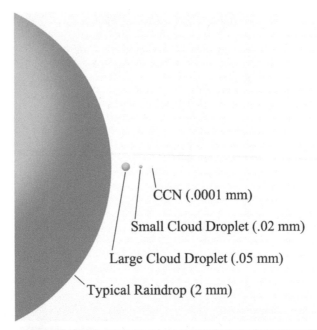

CCN (.0001 mm)

Small Cloud Droplet (.02 mm)

Large Cloud Droplet (.05 mm)

Typical Raindrop (2 mm)

Figure 5.2. Comparative diameters of raindrops, large cloud droplets, small cloud droplets, and cloud condensation nuclei (CCN) (not shown at actual size).

Clouds form when the atmosphere becomes saturated or supersaturated and condensation occurs. This leads to the creation of **cloud droplets**, which form on small, microscopic particles known as **cloud condensation nuclei**, or **CCN** (figure 5.2). CCN are vanishingly small and are composed of dust, clay, smoke, pollution, sea salt, or other small particles. There are literally hundreds or, more commonly, thousands of CCN per cubic inch of air, and as a result, hundreds or thousands of microscopic cloud droplets form per cubic inch.

You create your own cloud when you step outside on a cold morning and can see your breath. As you exhale, the water vapor in your breath (as well as a small amount from the atmosphere) condenses into thousands of tiny cloud droplets. This is an example of a **mixing cloud** because it forms from when the moisture-laden air in your breath mixes with the atmosphere.

The clouds that produce most snowstorms form not from mixing but from the cooling of moisture-laden air as it is lifted to higher altitude and lower pressure. Although not as fast and dramatic, this cooling is like the cooling of air as it exits a pressurized tire or aerosol can, and it occurs when air is forced to rise over a mountain range like the Wasatch Mountains.

Cloud Droplets

Snow and Ice Particles

Figure 5.3. Glaciating clouds over the Salt Lake Valley. Note the more fibrous appearance where the clouds consist primarily of snow and ice particles.

Glaciation

The second step, known as **glaciation**, begins when some of the cloud droplets freeze and become ice crystals (figure 5.3). This sounds simple, but it requires more than simply cooling the cloud to below what we usually think of as the freezing point (32°F). To freeze, water needs a particle known as an **ice nucleus**. An ice nucleus serves as a guide for water to arrange its molecules into an ice crystal. The best ice nucleus is ice itself, and it enables water to freeze at 32°F since it has a structure that precisely matches that of an ice crystal. In the absence of ice, however, a cloud droplet can become **supercooled** and remain unfrozen at temperatures below 32°F.

The temperature to which a supercooled cloud droplet must be cooled before it freezes depends on the structure of any particles it forms on or contacts. Naturally occurring particles that cause a cloud droplet to freeze at temperatures above about 20°F are extremely rare. As a result, clouds rarely

Figure 5.4. Rime on a chairlift at Brianhead ski area in southern Utah. Ice is deposited on the right (windward) side of the chairlift. Courtesy Meteorological Solutions, Inc. / Casey Lenhart.

glaciate unless they extend to altitudes where the temperature is below 20°F. Sub-freezing clouds that do not extend to altitudes where temperatures are below 20°F often produce rime in mountainous regions since the supercooled cloud droplets freeze when they collide with chairlifts, lift towers, trees, or snow on the ground (figure 5.4). Riming is usually only a minor annoyance for Utah skiers, but it can be a serious mountain weather hazard in other parts of the world.

In the 1940s, a scientist at General Electric, Bernard Vonnegut (brother of author Kurt Vonnegut), discovered that silver iodide smoke has a structure extremely close to that of ice and could act as an ice nucleus at temperatures as high as 25°F. This discovery formed the basis for **cloud seeding** (figure 5.5). The effectiveness of cloud seeding for precipitation enhancement remains controversial among meteorologists.

As the temperature decreases, water becomes less picky and will partner up with a wider range of particle types to freeze. The fate of an individual cloud droplet depends on its temperature and whether it contains or contacts a particle that can serve as an ice nucleus. Occasionally ice particles shatter and the resulting ice fragments collide with and freeze other cloud droplets.

This process, known as **ice multiplication**, further assists cloud glaciation. Nevertheless, there are so many cloud droplets and so few ice particles and nuclei that most clouds consist of a mixture of supercooled water and ice. In these **mixed-phase clouds**, there is usually more ice in the upper portion of the cloud, which is colder, and more supercooled water in the lower portion of the cloud, which is warmer (figure 5.6). Mixed-phase clouds produce most (but not quite all) of the snow that falls on Earth.

Vapor Deposition

Once we have a mixed-phase cloud, Mother Nature moves to step 3, **vapor deposition**. Vapor deposition is the dominant process in making the famed stellar dendrite and many other snowflake types and involves water vapor condensing directly into ice without passing through a liquid phase. This sounds strange, but frost forms from vapor deposition. The water vapor just skips dealing with the liquid phase.

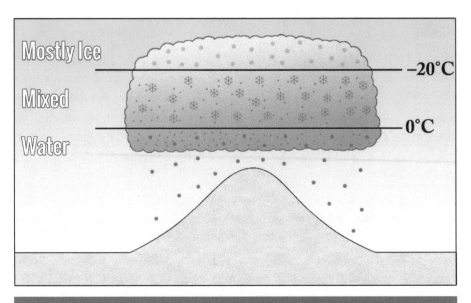

Figure 5.6. Mixed-phase clouds contain supercooled water and ice above the freezing level but may be mostly ice where temperatures are lower.

In a mixed-phase cloud, snowflakes grow from vapor deposition. Mother Nature has stacked the deck so this can occur. At temperatures below 32°F, the relative humidity for ice is higher than that for water. When the atmosphere is saturated for water, it is supersaturated for ice. As a result, in a mixed-phase cloud, the snowflakes grow, often at the expense of the cloud droplets (figure 5.7). This is known as the Wegener-Bergeron-Findeisen process (hereafter the **Bergeron process,** as it is commonly abbreviated) in honor of meteorologists Alfred Wegener, Tor Bergeron, and Walter Findeisen, who first explored this as a mechanism for snowflake growth.

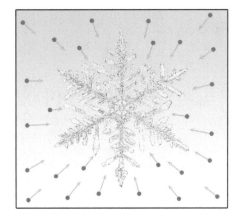

Figure 5.7. Owing to the difference in relative humidity for water and ice, snowflakes grow in a mixed-phase cloud at the "expense" of the cloud droplets.

Condensation, glaciation, and vapor deposition through the Bergeron process are essential steps in Mother Nature's snowmaking system. Without all three, you don't get a snowstorm. Sometimes, however, Mother Nature adds an additional step or two. Each of these steps has the potential to ruin a good powder day.

Riming

In some storms, Mother Nature adds riming to the mix. We've already discussed rime, which forms when supercooled cloud droplets collide with chairlifts, lift towers, trees, or the snow surface. The concept in a snowstorm is the same, except the cloud droplets freeze when they collide with snowflakes (figure 5.8). In some storms, there is little if any riming, but at the extreme, the riming can be so prolific that the original snowflake becomes entirely coated in frozen cloud droplets, forming graupel.

Riming increases the density of snowflakes by filling in the cavities and pores between the branches and arms. Snowstorms comprised of heavily rimed snowflakes sometimes create difficult skiing conditions, although graupel can produce surprisingly good skiing conditions since it isn't very sticky and creates a uniquely smooth and creamy snow consistency.

In mountainous regions, riming tends to be more prolific in warmer storms and in storms with strong updrafts. Warm storms typically have more supercooled water, while the updrafts keep snow and ice particles suspended while they are being rimed. In maritime mountain ranges like the Cascades, there are also fewer CCN, thanks to clean Pacific air. This leads to big cloud droplets that are extremely effective at riming and producing heavily rimed snow or graupel.

Aggregation

Finally, there is the fifth step, **aggregation**, which is the merger of two or more snowflakes. Aggregation most readily occurs with snowflakes that have arms and branches, which more easily become entangled, and at temperatures near or above 32°F when snow tends to be stickier. The resulting snowflakes are called **aggregates**. For big aggregates, it also helps if the wind is light, since strong winds often break up larger snowflakes.

PUTTING THE FIVE STEPS TOGETHER

We've discussed the five steps in sequential order, but in a real snowstorm, these processes are constantly operating together. A real snowstorm is a stew of cloud droplets, ice particles, and snowflakes that is constantly being mixed by winds and turbulence. Every snowflake has a unique life cycle. Let's take a closer look at what controls the shape, size, and composition of a snowflake and the quality of snow produced by a storm.

Figure 5.8. A falling snowflake can collect cloud droplets, which freeze on contact and produce snowflakes that may be lightly rimed (upper right), moderately rimed (center right), or completely coated to form graupel (lower right). Images courtesy Electron and Confocal Microscopy Laboratory, Agricultural Research Service, US Department of Agriculture.

Snowflake Habit

The shape of a snowflake, called a **habit**, varies depending on temperature and supersaturation (relative to ice). Precisely why snowflake habits change with temperature remains a bit of a mystery, but we have a pretty good handle on the conditions that produce certain types of snowflake habits.

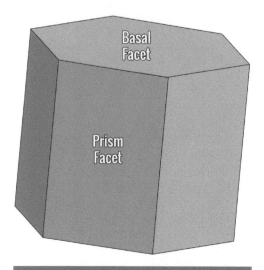

Figure 5.9. The prism and basal facets of an ice crystal.

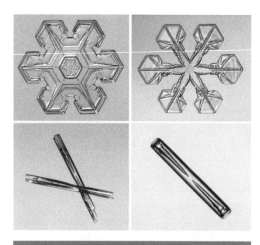

Figure 5.10. Images of a stellar plate (upper left), a sectored plate (upper right), needles (a type of column, lower left), and a hollow column (lower right). Courtesy Ken Libbrecht and snowcrystals.com.

The basic shape of an ice crystal is a **hexagonal prism**, which has eight faces, known as **facets** (figure 5.9). There are six prism facets, which form the hexagon sides, and two basal facets, which form the top and bottom. Snowflakes that have short prism facets and large basal facets are called **plates**, whereas snowflakes that have long prism facets and small basal facets are called **columns** (figure 5.10). Branches can form at the corners of the hexagonal prism, leading to more complex shapes.

Several diagrams have been developed to illustrate the dependence of habit on temperature and supersaturation for ice. For example, in the diagram developed by Japanese scientist Ukichiro Nakaya, plates form at temperatures between 0°C and –4°C, columns between –4°C and –10°C, and plates again between –10°C and about –20°C (figure 5.11). This diagram has the advantage of being relatively straightforward, but the real world is a complicated place.

The snowflakes that fall during winter storms have formed and grown at a variety of altitudes and temperatures. This usually leads to a diverse mix of habits, including some that are highly irregular and do not resemble the beautiful, symmetric crystals found in habit diagrams. In addition, most snowflakes that fall during a storm are defective, even in Utah. They are often poorly formed or have broken into pieces while colliding with other snowflakes. They are frequently rimed or aggregated (figure 5.12). Although I enjoy examining snowflakes and trying to identify the dominant habits, sometimes it's best just to ski it if it's white.

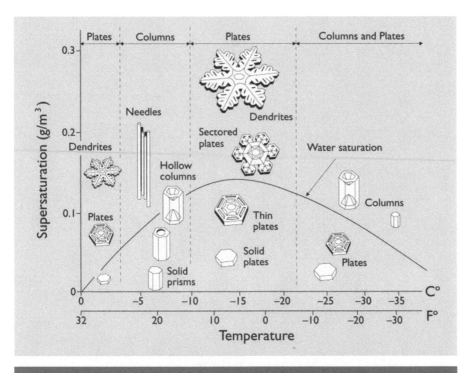

Figure 5.11. Snowflake habits. Courtesy Ken Libbrecht and snowcrystals.com. Adapted from a diagram by Furakawa, with habits based on work by Japanese scientist Ukichiro Nakaya.

Blower Pow and Wild Snow

Every now and then, Mother Nature produces a big storm consisting primarily of lightly rimed stellar and spatial dendrites.[1] Dendrites are special because they contain lots of branches and arms with cavities and pores that are full of air rather than ice. Fresh snowfalls that consist primarily of lightly rimed or unrimed dendrites can have incredibly low water content, sometimes as low as 2 percent, if the winds are light. Skiers call such dry snow "blower pow" or "cold smoke," but meteorologists call it **wild snow**.

Wild snow has a water content of 4 percent or less. Many storms in the Cottonwood Canyons conclude with a period of wild snow, but storms that consist entirely of wild snow are rare. Only 6 percent of the storm days at Alta qualify as wild snow events, and nearly all occur in December, January, or February. One of Alta's most extreme wild snow events occurred on February 7, 1990, when twenty-six inches of 2.5 percent water content snow fell.

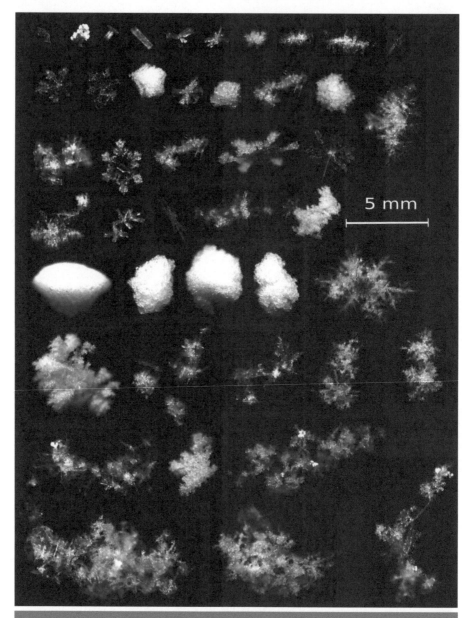

Figure 5.12. Snowflake images taken by a Multi-Angle Snowflake Camera (MASC) during winter storms at Alta. Courtesy Tim Garrett and Cale Fallgatter, University of Utah.

A wild snow event requires several ingredients to come together. Nearly all the snowflakes must grow in what meteorologists call the **dendritic growth region**. This is a region of the atmosphere where the temperature is between about –12°C and –18°C, and if the supersaturation for ice is high, as it frequently is, the favored snowflake habit is the dendrite (figure 5.11). At Alta, most of the snowflakes during wild snow events are growing at altitudes near or just above the ski area, which means the air temperature at the ski area will probably be in or just above the –12°C to –18°C range. In addition, the rising motion responsible for precipitation generation must be just right to enable the dendrites to grow and fall without a lot of riming. Finally, the winds also need to be light so that the dendrites don't collide and break up in the air or on the ground. It's unusual to have an extended period when all these ingredients come together, which is why wild snow events are rare. On the other hand, it's not unusual for Mother Nature to lay down a few inches of wild snow at the end of a storm and, as we learned in chapter 1, that produces a right-side-up snowfall that's perfect for powder skiing (figure 5.13).

Sierra Cement, Utah Style

Don't be fooled by the ads. It's not always a ski paradise in Utah. We sometimes have warm storms with high snow levels that can approach or even exceed 8,000 feet. Riming, aggregation, and melting during these storms produces snow that better resembles Sierra cement or Cascade concrete than the Greatest Snow on Earth, with a water content that can reach or exceed 20 percent.

The moisture that fuels many of the storms that bring Sierra-like conditions to Utah usually originates over the tropical Pacific Ocean, sometimes near Hawaii. It flows to Utah in a narrow stream known as an atmospheric river. An example is the December 18–20, 2010, atmospheric river that brought prolific snows to Mammoth Mountain in California before setting its sights on Utah (see chapter 3 and figure 3.5). As this atmospheric river penetrated into Utah, it capped the end of a storm cycle that produced 75 inches of snow and 12.33 inches of snow water equivalent at a mid-mountain observing site at Sundance ski area from December 17–23, 2010. This precipitation fell as heavy wet snow or even rain at times below about 7,500 feet. Sundance was forced to close on December 19 due to high avalanche hazard, and massive avalanches on Mount Timpanogos ran over 3,000 vertical feet, leaving large debris piles (figure 5.14). Wild things happen when the tropics come to Utah.

The Graupel Storm

Sometimes warm storms produce lots of graupel. During these storms, the skiing is often surprisingly good. Graupel is not very sticky. It lacks cohesion.

Figure 5.13. Caroline Gleich shreds right-side-up powder topped by low-density dendrites at Alta. © Lee Cohen. Used with permission.

Figure 5.14. Flooding near Sundance and avalanches on Mount Timpanogos during the December 2010 atmospheric river event. Courtesy Bill Nalli.

Graupel has a high water content and stings when it hits your face, but it skis remarkably well. In fact, graupel is a real treat for discriminating Utah skiers who know a good thing when they ski it. Although graupel is most common in warm storms, it can be produced in colder storms if there are strong updrafts, as found in wintertime thunderstorms.

ARTIFICIAL SNOW

Grand Targhee ski area in Wyoming's Teton Range has one of the best slogans in the ski business: "Snow from heaven, not hoses." Grand Targhee certainly has every right to boast about its snow. The snow climate at Grand Targhee is one of the best in the world and is comparable to that found in Utah's Cottonwood Canyons.

I love the Grand Targhee slogan because there really is no comparison between snow from heaven and snow from hoses. Mother Nature makes great snow with vapor deposition. Humans make it by blowing water droplets from guns. As a result, artificial snow is composed primarily of frozen water droplets (figure 5.15). Because artificial snow is just frozen water droplets, it might better be named artificial sleet.

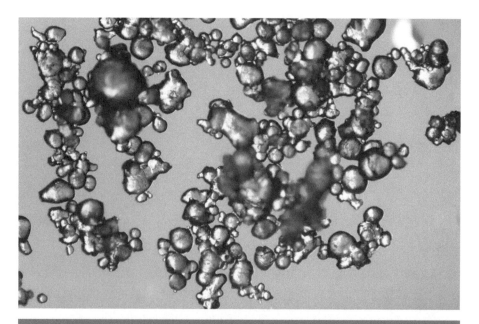

Figure 5.15. Close-up photo of artificial snow. Courtesy Ken Libbrecht and snowcrystals.com.

As much as I despise artificial snow, it is critical for the Utah ski industry, even at a ski area like Alta and especially at less snowy ski areas on the Park City side of the range. Meteorologists have a saying: "Climate is what you expect, weather is what you get." Although Utah has a great snow climate, the weather during November and December is not always snowy. Artificial snow provides a necessary supplement to that provided by Mother Nature during some critical holiday seasons.

NOTE

1. The spatial dendrite is a deviant version of the stellar dendrite with one or more arms that extend out of the main plane of the snowflake.

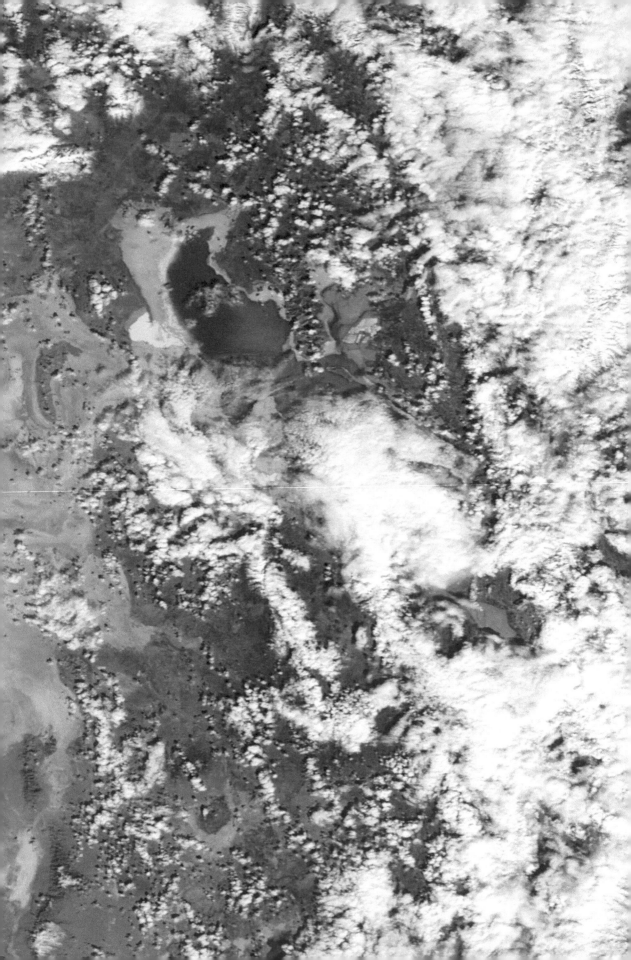

6 Lake Effect

Conventional wisdom and marketing hype suggest that the Great Salt Lake effect is the holy grail of Utah powder skiing. For example, skisaltlake.com claims that "storms suck up moisture as they pass over the nearby Great Salt Lake and [drop it] on Salt Lake's Mountains just miles away. The light powder snow, thanks to the lake's salinity, falls en masse upon Alta, Brighton, Snowbird, and Solitude, creating some of the best powder skiing and snowboarding in the world."

On the other hand, meteorologists have a love-hate relationship with the Great Salt Lake effect. Many Utah meteorologists are skiers and love a powder day as much as anyone, but forecasting the Great Salt Lake effect is incredibly difficult. In fact, it is so difficult that meteorologists call it the **dreaded lake effect** (DLE), although we use saltier language in private.

What is the Great Salt Lake effect and why is it so hard to forecast? To answer these questions, we first need to learn about the Great Salt Lake, one of the most unusual bodies of water in the world.

THE GREAT SALT LAKE

The Great Salt Lake is the largest body of water in the western United States and the fourth-largest **terminal lake** in the world (figure 6.1). A terminal lake has no outlet, so the only escape for water that flows into the Great Salt Lake is evaporation. As a result, the depth, surface elevation, and area of the Great Salt Lake vary significantly, increasing during snowy years that produce a lot of spring runoff and decreasing during droughts (figure 6.2). Geological records suggest that over the past 10,000 years, the Great Salt Lake has never gone completely dry and has crested twice near or above 4,217 feet. At that level, water begins to flow through a low divide to the Great Salt Lake Desert.

Since the mid-1800s, the surface elevation of the Great Salt Lake at the Saltair Boat Harbor on the south shore has fluctuated between a high of just over 4,111 feet in 1872, 1873, and 1986 and a low of just under 4,191 feet in 2021 (as I write this, the forecast for the summer of 2022 is for the Great Salt Lake to reach an all-time low near 4,189 feet). Because much of the land around the Great Salt Lake is very flat, these changes in surface elevation produce large changes in lake area from as high as 3,330 square miles in 1986 to as low as 950 square miles in 2021. Nevertheless, even at its largest, the Great Salt Lake is much smaller than Lake Ontario, which at 7,320 square miles is the smallest of the eastern Great Lakes. Fortunately, the Great Salt Lake is longest (about 70 miles) from northwest to southeast, which means that cold northwest-

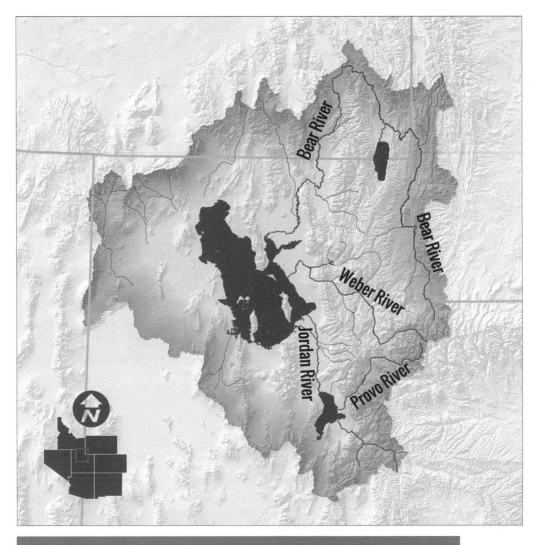

Figure 6.1. The closed drainage basin of the Great Salt Lake.

erly flow remains over the lake as long as possible before moving over the Cottonwood Canyons.

The Great Salt Lake is also remarkably shallow. At a surface elevation of 4,200 feet it has an average depth of 16 feet and a maximum depth of 35 feet. You could think of it as a massive puddle rather than a lake. Because it is so shallow, the Great Salt Lake warms and cools very rapidly. Although deeper lakes like the Great Lakes take a long time to warm up and don't produce much lake effect in the spring, the Great Salt Lake is primed and ready to go any time of year after a week of warm weather.

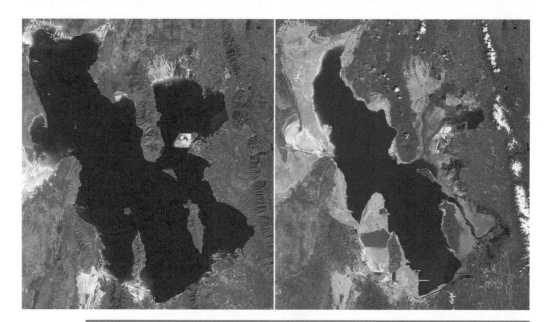

Figure 6.2. Landsat satellite images of the Great Salt Lake based on data collected in September 1987 (left) and April and May 2021 (right). Images courtesy of the USGS.

The Great Salt Lake is one of the saltiest bodies of water in the world, with a **salinity** (salt content) near 12 percent in its southern half and 27 percent in its northern half (3.5 and 8 times saltier than the ocean, respectively). This salinity contrast occurs because a rock-fill railroad causeway divides the lake and limits the flow of water between the two halves. The salinity is lower in the southern half because it receives most of the fresh-water inflow. The two halves even have different colors due to differences in the concentration of algae and bacteria that thrive in high-salinity environments (figure 6.3).

Salinity affects the Great Salt Lake effect in two ways. First, salt lowers the freezing point of water. As a result, the Great Salt Lake never freezes except near freshwater inlets. This is good for lake effect because the lake can always interact directly with the atmosphere. Second, salt water does not evaporate as quickly as freshwater. In fact, evaporation over the northern half of the lake is probably 30–40 percent lower than it would be if it were freshwater. Unfortunately, this is not good for lake effect because it reduces the rate at which moisture from the Great Salt Lake is pumped into the atmosphere.

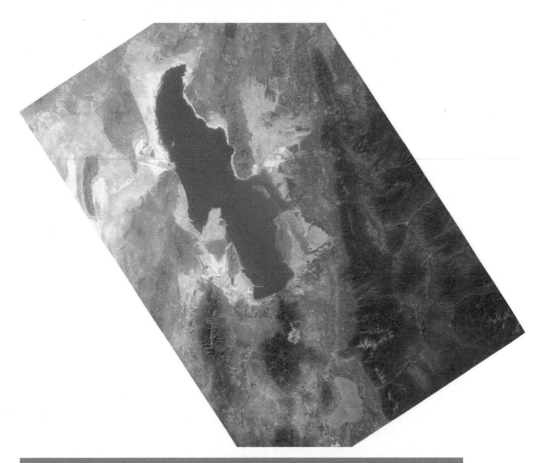

Figure 6.3. Photo of the Great Salt Lake taken from the International Space Station on August 19, 2003, showing the color contrast between the northern and southern halves. Mission-roll-frame ISS007-E-13002 courtesy of the Image Science and Analysis Laboratory, NASA Johnson Space Center. Available at http://eol.jsc.nasa.gov.

LAKE-EFFECT INGREDIENTS, TYPES, AND TIMING

Great Salt Lake–effect snowstorms occur when relatively cold westerly, north-westerly, or northerly flow moves over the Great Salt Lake. However, it takes more than just cold air to produce a lake-effect snowstorm. There are two other key ingredients: upstream moisture and a trigger.

When cold air moves over a relatively warm lake, heat and moisture are pumped into the atmosphere, but the Great Salt Lake is relatively small (compared to the Great Lakes) and super salty, which means it can't pump enough moisture into the atmosphere to generate a storm if the upstream air mass is

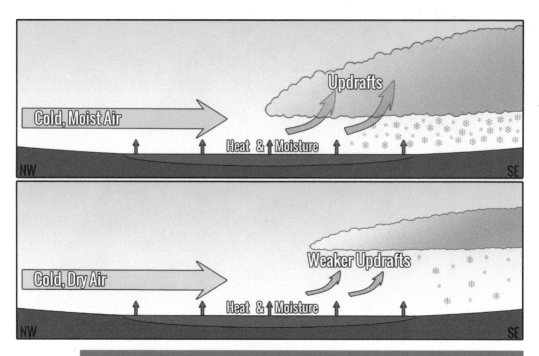

Figure 6.4. The development and intensity of lake-effect precipitation is influenced by the characteristics of the upstream air mass. A moist upstream air mass (top) produces a stronger, more intense lake-effect system than a dry upstream air mass (bottom).

bone dry (figure 6.4). In most Great Salt Lake–effect snowstorms, the relative humidity of the upstream air mass is at least 60 percent. In addition, modeling studies of Great Salt Lake–effect snowstorms show that a 10 percent increase in the upstream relative humidity can double the lake-effect snowfall. If you want a lake-effect storm, it's best to start with some upstream moisture.

But cold, moist air is still not enough. Lake effect needs a trigger. The lake-effect trigger is something meteorologists call **convergence**. When airstreams converge over the Great Salt Lake, the air has nowhere to go but up, and the resulting lift triggers and helps organize lake-effect storms.

The Great Salt Lake can generate its own trigger. After sunset, the air surrounding the lake cools faster than the air over the warm lake surface. This produces **land breezes**, currents of air that blow from the surrounding land and converge over the Great Salt Lake (figure 6.5). This **land-breeze convergence** often serves as the trigger for lake-effect storms, including intense, isolated **mid-lake bands** that can have snowfall rates near three inches per hour (figure 6.6). I love skiing in mid-lake bands, but they occur only about 20 percent of the

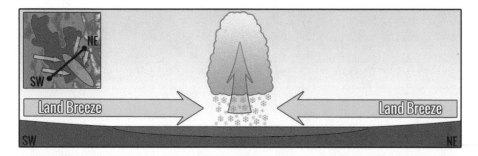

Figure 6.5. Triggering of a lake-effect storm by land-breeze convergence.

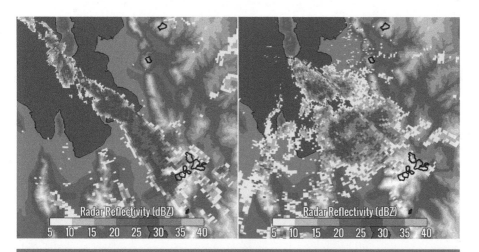

Figure 6.6. Examples of a mid-lake band at 4:30 a.m. on November 23, 2001 (left), and broad-area coverage at 11:27 p.m. on November 28, 2006 (right). In the left-hand image, echoes over the high terrain around the Cottonwood Canyons were removed due to the clutter produced by the beam intersecting the ground.

time during lake-effect storms. Most lake-effect storms are less concentrated and cover a broader area with less intense snowfall.

Land breezes typically reach maximum intensity overnight or early in the morning and weaken during the day. As a result, lake-effect snow tends to be more frequent and intense at night and in the morning than in the afternoon. What could be better for powder skiing than that?

Recent research indicates that there is another mechanism for initiating or enhancing lake-effect storms. Sometimes when there is flow from the northwest, the low-level flow is funneled into the Salt Lake Valley by the mountains

Figure 6.7. Triggering or enhancement of lake-effect precipitation as the mountains south and east of the Great Salt Lake funnel flow into the Salt Lake Valley.

to the south and east of the Great Salt Lake (figure 6.7). This **terrain-driven convergence** can trigger or enhance some lake-effect storms and may also help strengthen some northwesterly flow storms that don't contain lake effect.

WHO GETS THE LAKE-EFFECT GOODS?

Given the orientation of the Great Salt Lake and the tendency for cold air to move into northern Utah from the northwest, lake-effect precipitation is greatest to the south and east of the Great Salt Lake (figure 2.15). This puts the Cottonwoods, including Alta, Snowbird, Brighton, and Solitude, squarely in the lake-effect crosshairs.

On average, lake-effect periods generate about 5 percent of the snow that falls in the Cottonwoods each winter. Most skiers and snowboarders are surprised the percentage isn't higher, but marketers and industry promoters suffer from delusions when it comes to the Great Salt Lake effect.

Considerably less lake-effect snow falls at the Wasatch Back ski areas, namely Park City Mountain Resort, Deer Valley, and Mayflower. During lake-effect storms, the Cottonwoods can get plastered while little or no snow falls in downtown Park City. This is because lake-effect storms are typically quite shallow and don't penetrate far across the Wasatch crest. In fact, even in the Cottonwoods, it is not uncommon to see filtered sun through the clouds and falling snow during lake-effect storms.

Less lake-effect snow also falls at the northern Wasatch ski areas of Snowbasin and Powder Mountain. This is related to meteorology and lake geometry. For these ski areas to be downstream of the Great Salt Lake, the flow needs to be from the west or southwest. There are fewer cold-air intrusions from these directions, and such flows move along the short axis of the lake rather than the long axis.

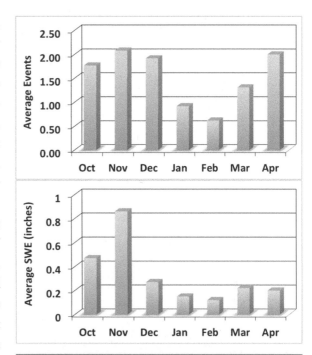

Figure 6.8. Average number of lake-effect events (top) and average amount of precipitation (snow water equivalent [SWE]) by month during lake-effect periods at Snowbird (bottom). Source: Alcott et al. 2012; Yeager et al. 2013.

WHEN TO GET THE LAKE-EFFECT GOODS

On average, lake-effect storms are most common and produce the most precipitation from October to December (figure 6.8). There is a midwinter lull that bottoms out in early February, followed by a second peak in March and April. There are, however, great variations from month to month and year to year. In a big lake-effect year like 2001–2002, lake-effect periods contributed about 12 percent of the precipitation that fell in the Cottonwood Canyons. On the other hand, during the 2002–2003 winter that followed, which featured a persistent storm track from the southwest, lake-effect periods produced only 1 percent of the precipitation.

EPIC LAKE-EFFECT STORMS

Most lake-effect periods are brief and don't produce large accumulations. Instead, they usually provide a few inches of low-density snow for face shots and chin ticklers at the end of a storm cycle. There are, however, some epic lake-effect storms that shall forever remain a part of Wasatch lore.

The 2004–2005 ski season started with an incredible late-October storm cycle that laid down 122 inches of snow at Alta. Remarkable backcountry skiing was had throughout the Wasatch Mountains, and Brighton opened for lift-served skiing on October 29. Capping it all off was a Halloween lake-effect storm that served up some fantastic trick or treating (figure 6.9). Laying down about two feet of cold smoke, the storm produced over-the-head powder skiing that grizzled Wasatch backcountry skier and National Weather Service meteorologist Larry Dunn called the Best Day of October Ever, or BDOE.

WITHER THE GREAT SALT LAKE?

Since about 2000, the southwest United States and northern Utah have been in the grips of a **megadrought**, a period of severe dryness lasting for at least a decade. Tree rings and other records suggest that the southwest United States has experienced extended periods of drought and wetness for periods of a decade or more in the past, with the Great Salt Lake rising and falling in response to these climate variations.

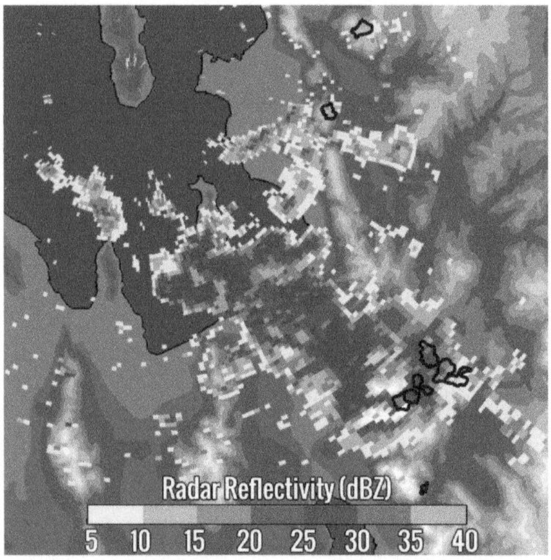

Figure 6.9. The 9:24 a.m. radar image of the lake-effect storm that produced epic powder skiing on October 31, 2004. Echoes over the high terrain around the Cottonwood Canyons were removed due to the clutter produced by the beam intersecting the ground.

In recent decades, however, two things have changed. First, we have been diverting water from the rivers that provide inflow to the Great Salt Lake. Recent research indicates that this has lowered lake level by about eleven feet. Second, we are experiencing a warming climate, which has turned what would have been a moderate megadrought into a severe one.

In 2021, the Great Salt Lake reached its lowest surface elevation on record at the Saltair Boat Harbor. As I write this, it is expected to drop to a new record low, perhaps 4,189 feet, during the summer of 2022. Perhaps by the time you read this we've had a few wetter years and the lake has risen some, or we've seen continued drought and further lower-

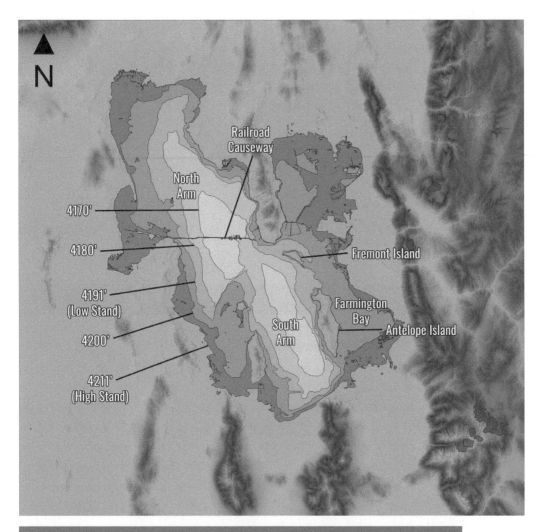

Figure 6.10. Shorelines of the Great Salt Lake at different surface elevations. Great Salt Lake bathymetry source: Tarboton (2017). Background hillshade sources: Esri, USGS, FAO, NOAA.

ing of lake levels. Regardless, many terminal lakes around the world, such as the Aral Sea and Owens Lake, have dried up due to water diversion, and now we have global warming as a drought enhancer. What will happen if the Great Salt Lake continues to wither?

As lake levels lower, the Great Salt Lake will become increasingly confined to a narrow area along the current major lake axis where the bottom of the lake is deepest (figure 6.10). At the 2021 low stand near 4,191 feet, very little water remained in Farmington Bay, and Antelope and Fremont Islands were no longer islands. At 4,180 feet, water is confined to the north and south arms, with the Great Salt Lake more of a Great Salt Finger Lake. At 4,170 feet, the lake is divided into two small pieces, one entirely in the south arm, the other centered on the railroad causeway.

As the Great Salt Lake shrinks, it is likely that the Great Salt Lake effect will also decrease. This won't be a catastrophe for skiing since the Great Salt Lake effect produces only about 5 percent of the total snowfall in the Cottonwoods, but it will add another insult to the injury of global warming, which is already having an impact on snow and snowpack. Additionally, as more and more of the lake bed is exposed, it will become a source of dust and, in addition to health impacts, dust is also a threat to Utah's snowpack. We will explore these issues in greater detail in chapters 10 and 11.

LAKE-EFFECT MYTHOLOGY

No Wasatch weather phenomenon is more misunderstood than the Great Salt Lake effect. For years it has been exploited by marketing hounds and misdiagnosed by the media. There are, however, three nuggets of misinformation that rise above the rest to stand on my Olympic podium of lake-effect myths.

Moisture sucked from the Great Salt Lake causes lake-effect storms. As we have discussed, there are three critical ingredients for lake effect: (1) cold air moving over a warm lake; (2) upstream moisture; and (3) a trigger. Take any of these out of the recipe and you'll be zooming the groomers rather than skiing the steep and deep. Moisture from the Great Salt Lake doesn't cause lake effect, but it is a lake-effect enhancer. In computer models, we can take moisture from the lake out and still get lake-effect snow, but the coverage and intensity of the snowfall is reduced.

Salt from the Great Salt Lake seeds storms, causing unusually dry snow. As we have seen in chapters 1 and 4, the water content of snow is determined by several factors, with temperature playing the greatest role. Lake effect occurs in cold air and frequently at temperatures that favor the formation of dry snow. This is the primary reason why many lake-effect storms produce great powder. Although there are traces of salt in the Wasatch snowpack, it comes primarily from dust storms, not snowstorms.

On the other hand, not all lake-effect storms produce low-density snow. There are some that produce large amounts of graupel, which skis great but has a high water content. These are usually warmer, highly unstable lake-effect storms with intense updrafts.

Size matters. The size of the Great Salt Lake fluctuates from year to year, so the idea here is that there's more lake effect in seasons when the lake is big. This myth has some truth to it. All else being equal, a bigger lake will produce a more intense lake-effect storm. The problem is that all else is never equal. The primary factor determining how much lake effect falls each season is not the size of the lake but the meteorology. For example, a season with frequent surges of cold air from the northwest will produce more lake-effect snow than one dominated by a warmer, southwesterly storm track. Nevertheless, if the Great Salt Lake continues to shrink in size, there is going to be a significant decline in lake-effect snowfall even during seasons with lots of northwesterly cold-air surges.

7 Alta Goes to War

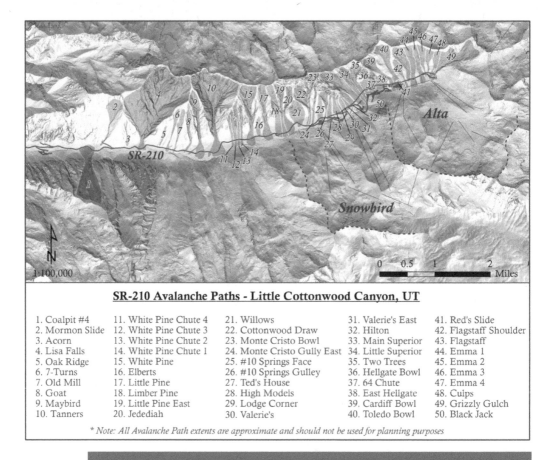

SR-210 Avalanche Paths - Little Cottonwood Canyon, UT

1. Coalpit #4	11. White Pine Chute 4	21. Willows	31. Valerie's East	41. Red's Slide
2. Mormon Slide	12. White Pine Chute 3	22. Cottonwood Draw	32. Hilton	42. Flagstaff Shoulder
3. Acorn	13. White Pine Chute 2	23. Monte Cristo Bowl	33. Main Superior	43. Flagstaff
4. Lisa Falls	14. White Pine Chute 1	24. Monte Cristo Gully East	34. Little Superior	44. Emma 1
5. Oak Ridge	15. White Pine	25. #10 Springs Face	35. Two Trees	45. Emma 2
6. 7-Turns	16. Elberts	26. #10 Springs Gulley	36. Hellgate Bowl	46. Emma 3
7. Old Mill	17. Little Pine	27. Ted's House	37. 64 Chute	47. Emma 4
8. Goat	18. Limber Pine	28. High Models	38. East Hellgate	48. Culps
9. Maybird	19. Little Pine East	29. Lodge Corner	39. Cardiff Bowl	49. Grizzly Gulch
10. Tanners	20. Jedediah	30. Valerie's	40. Toledo Bowl	50. Black Jack

* Note: All Avalanche Path extents are approximate and should not be used for planning purposes

Figure 7.1. Avalanche paths that intersect SR-210 and other roads and parking lots in Little Cottonwood Canyon. Courtesy Utah Department of Transportation / Adam Naisbitt.

All major ski areas and most mountain highways in the Wasatch Mountains are susceptible to avalanches, but Little Cottonwood Canyon is an especially dangerous place. Roughly fifty **avalanche paths** threaten State Route 210 (SR-210) and other roads and parking lots in Little Cottonwood Canyon (figure 7.1), which are hit by an average of thirty-three avalanches per year (figure 7.2). The combination of frequent avalanches and heavy traffic leads to the highest avalanche hazard of any highway in the United States. Seven major avalanche paths lie above the town of Alta, while others cross parking lots and portions of Snowbird village. Add in thousands of acres of avalanche terrain within the ski area boundaries, and it takes a Herculean effort by many snow-safety professionals just so you can get the goods after a storm. Those who came to Little Cottonwood Canyon in years past were not so lucky.

Figure 7.2. An avalanche triggered by artillery on Mount Superior above the Little Cottonwood Canyon highway, which was closed to vehicle traffic. Courtesy Adam Naisbitt.

THE MINING ERA

It was not long after Mormon pioneers arrived in Utah in 1847 that development came to Little Cottonwood Canyon. First came logging and lumber mills, then mining. By 1873, several mining camps were operating in the canyon, and Alta was a town with 180 buildings and a seasonal population of about 3,000. Mining is dangerous enough, but now add the natural hazards of working in a steeply sloped mountain canyon that receives more than 500 inches of snow a year. Further, while the south-facing slopes north of Alta are naturally treeless, logging for buildings and mining denuded the north-facing slopes above Alta, exposing more terrain to the threat of avalanches (figure 7.3). Avalanche disasters would strike and strike often.

The number of people killed in avalanches during Alta's mining days is unclear. In 1884, the *Salt Lake Tribune* reported, "Since mines at Alta were first opened 14 years ago, 143 people have been killed by snowslides in and around Alta" (Kalitowski 1988). According to the Alta Historical Society, documented

Figure 7.3. C. R. Savage photo of Alta's "Rustler Mountain" on July 3, 1885. Ski runs in this area today include High Nowhere, North Rustler, Eagle's Nest, and High Rustler. Courtesy Utah State Historical Society.

fatalities total seventy-four. In any event, by 1887 the mining boom was over, and thanks largely to avalanches, there was only one building left standing at Alta. Mother Nature had no sympathy for those exploiting an avalanche-raked mountain canyon during the 1800s.

SKIING COMES TO ALTA

Eventually people realized that the real treasure in Little Cottonwood Canyon wasn't silver but powder. The "original" 1930s Utah Interconnect involved taking a train to Park City, skiing through Guardsman Pass to Brighton, overnighting at the Rose Hotel, traveling to Alta via Catherine Pass, and then exiting out Little Cottonwood Canyon. With no lifts, it was quite an adventure.

Alta first opened for lift-served skiing in 1938 with a rickety chairlift constructed partly from an old mining tramway. Nevertheless, the specter of avalanches remained. In 1939, Wasatch National Forest supervisor James E. Gurr, a strong proponent of winter-sports development in the Cottonwood Canyons,

Figure 7.4. Sverre Engen getting the goods in Sun Valley, Idaho. Courtesy Special Collections Department, J. Willard Marriott Library, University of Utah.

expressed concern: "Winter sports enthusiasts should appreciate the fact that there is some hazard connected with the use of snow-clad slopes and should consistently practice such safety measures as are necessary to minimize these hazards. If they will do this and will cooperate in following our suggested safety measures, we can and will continue to use our outstanding and valuable winter sports areas" (Kalitowski 1988).

Gurr and district ranger W. E. Tangren established snow-safety strategies for Alta, with Tangren becoming the first Forest Service avalanche observer during the 1938–1939 ski season. In late 1939, Gurr hired Sverre Engen to assist Tangren. Sverre, along with his brother Alf, was a famous ski jumper and would eventually become a powder-skiing icon (figure 7.4). In the early 1940s, he became

January 12–17, 1881: Avalanches during a period described by Alta resident Dr. F. J. Simmons as "one of the most eventful known in the history of Alta" leave an estimated fifteen people dead (Kalitowski 1988).

March 7, 1884: Avalanches pummel Alta, killing twelve.

February 13, 1885: Avalanches take another fifteen lives at Alta. Fifty feet of snow deposited on what was then Main Street.

February 1939: First use of explosives for avalanche mitigation in Little Cottonwood Canyon, with the resulting slide burying 2,000 feet of the highway under fourteen feet of snow.

December 1972: A huge avalanche fills the top floor of the Alta Lodge with snow and leaves a Volkswagen Beetle on the roof.

December 29, 1973: Avalanches cause $250,000 in damage (1973 dollars) to vehicles and buildings at Alta and Snowbird.

Spring 1983: The original Our Lady of the Snows Chapel is damaged by a wet spring avalanche and torn down. The current, more heavily fortified chapel was built in 1993–1994.

March 14, 2002: Avalanche mitigation work releases a slide that crashes into the Alta-Peruvian Lodge, knocking out windows in some rooms and burying several cars. Guests were safely lodged elsewhere at the time.

February 11–18, 2021: A 103-inch storm cycle results in a sixty-hour interlodge at Alta and numerous long-running avalanches depositing deep snow and debris on State Route 210 in Little Cottonwood Canyon.

the Forest Service's first "snow ranger," charged with collecting snow and weather data, developing a snow-safety plan for Alta, and opening and closing the Little Cottonwood highway.

Snow safety at the time was rudimentary, based on passive measures such as closing the highway and ski terrain in and beneath slide-prone areas. Skiers wouldn't stand for this today and neither did the powder hounds of the 1940s. In 1946, Forest Service supervisor Felix Koziol observed, "Winter skiers wander all over the landscape. Many are not content to stay on the practice slopes and on the ski runs close to the lifts. An increasing number are seeking ways of getting far out into the alpine hinterlands, far away from the snow bunny crowd" (*Alta Powder News* 2009).

Before 1946, the use of explosives for avalanche mitigation was limited and experimental. But that was about to change: Alta was going to war.

THE MONTY ATWATER ERA

In 1945, Sverre Engen became director of the Alta ski school, and Monty Atwater was hired as snow ranger (figure 7.5). Atwater, a veteran of the Tenth Mountain Division, had a keen knowledge of explosives. Under his leadership, the use of explosives and artillery to mitigate the avalanche hazard in Little Cottonwood Canyon shifted into overdrive.

In 1949, Atwater and Koziol convinced the Utah National Guard to use a seventy-five-millimeter French howitzer for avalanche mitigation in Little Cottonwood Canyon (figure 7.6). On March 30, fifteen rounds were fired into avalanche paths, proving the concept was viable. It

Figure 7.5. Monty Atwater. Courtesy Special Collections Department, J. Willard Marriott Library, University of Utah.

required a long period of negotiation, however, to persuade the military that such weapons should be used for civilian purposes. At first, only members of the National Guard were authorized to shoot the howitzer, but this proved wholly impractical since often they couldn't get up the canyon after storms. Atwater began to shoot the weapon covertly, but he was eventually given permission to use it whenever necessary. He and his "Avalanche Hunters" were in business.

Figure 7.6. Ed LaChapelle photograph of Monty Atwater using a French howitzer for avalanche mitigation at Alta in the early 1950s. Courtesy Special Collections Department, J. Willard Marriott Library, University of Utah.

THE ED LACHAPELLE ERA

Monty Atwater was a talented individual and is widely recognized as the "Father of Snow Avalanche Work in the United States," but he wasn't a trained scientist (he held a bachelor's degree in English literature). In 1952, he brought Ed LaChapelle to Alta. LaChapelle held degrees in physics and math from the University of Puget Sound and conducted his postgraduate work at the Swiss Federal Institute for Snow and Avalanche Research in Davos, Switzerland, the leading center for avalanche research in the world. He was an inventive and methodical scientist who established a **snow-study plot** (now de rigueur

Figure 7.7. Ed LaChapelle at the Atwater snow study site, Alta, Utah. Courtesy Special Collections Department, J. Willard Marriott Library, University of Utah.

for snow-safety organizations) and advanced Atwater's avalanche-mitigation methods (figure 7.7). LaChapelle also wrote three books that remain essential reading for avalanche and mountain-weather geeks today: *The ABCs of Avalanche Safety* (the most recent edition, 2003, coauthored with the late Sue

Ferguson, an avalanche pioneer in her own right), *Field Guide to Snow Crystals* (1969), and *Secrets of the Snow: Visual Clues to Avalanche and Ski Conditions* (2001).

After Monty Atwater's departure to what is known today as Palisades Tahoe to lead avalanche-mitigation efforts for the 1960 Winter Olympics, LaChapelle pushed avalanche research at Alta to new heights before he left for a faculty position at the University of Washington in 1967. In the late 1960s, the Forest Service elected to phase out avalanche research at Alta and shifted avalanche-mitigation responsibilities in Little Cottonwood Canyon to Alta ski area and the Utah Department of Transportation, with Snowbird joining the mix during its inaugural 1971–1972 season.

THE MODERN ERA

There is a tremendous spirit of cooperation in Little Cottonwood Canyon, and avalanche mitigation today involves an allied effort between groups and individuals at the Utah Department of Transportation (UDOT), Alta ski area, Snowbird, the Wasatch Powderbird Guides, the Town of Alta, the US Forest Service, and the Unified Police Department of Greater Salt Lake.

UDOT's primary responsibility is avalanche safety along the Little Cottonwood Canyon Road, also known as SR-210. Substantial infrastructure is used to monitor and mitigate avalanches along the road, including:

- 105-millimeter howitzers at Alta (operated by Alta ski area) and Snowbird (operated by Snowbird) that collectively fired an average of about 600 rounds per year into the south-facing avalanche slopes on the north side of the highway through the mid 2010s (figure 7.8), although this number is now declining as other avalanche-mitigation systems are deployed.
- **Wyssen avalanche towers** that lower and detonate explosive charges over the snowpack (figure 7.9).
- The Wasatch Powderbird Guides helicopter, which is used for heli-bombing.
- **Gazex avalanche-mitigation systems** that direct a blast of air generated from propane combustion onto the snowpack (figure 7.10).
- An infrasonic avalanche detection system for monitoring avalanche activity in problematic mid-canyon avalanche paths.
- Numerous weather stations operated by UDOT, Alta ski area, Snowbird, and other groups to monitor wind, precipitation, and temperature within the canyon.

Figure 7.8. Todd Greenfield and Paul Garskey at the Valley Howitzer at Snowbird. Courtesy Adam Naisbitt.

Monitoring snow and weather conditions along the highway is a 24/7 operation involving a supervisor, snow-safety personnel, an on-site meteorologist, and additional meteorological support provided by the National Weather Service forecast office in Salt Lake City. Strategies for avalanche mitigation are crafted each afternoon for the following night and day. When necessary, the road is closed in the morning for mitigation work, with the goal of having it open for traffic by 8:30 a.m. Longer closures occur, however, when the avalanche hazard remains a concern or if considerable avalanche debris must be removed from the road. In some situations, the road is closed during the day, such as when the avalanche hazard increases rapidly due to high snowfall rates or rapid warming. If the avalanche hazard increases rapidly at night, the road is simply closed until morning since daylight is required for most mitigation work. All this effort greatly reduces but does not eliminate avalanche risk along the highway (figure 7.11).

Vital to opening the highway and ski areas are avalanche-mitigation activities at Alta and Snowbird. Snow-safety supervisors at both ski areas are early

Figure 7.9. A Wyssen avalanche tower above the town of Alta in Little Cottonwood Canyon. Charges in the round box at the top of the tower can be lowered remotely and detonated over the snow surface to trigger avalanches. Courtesy Mike Kok.

risers, frequently up at 3:30 a.m. Their staffs assemble before sunup, eat breakfast, and are on the lifts as early as 6:00 a.m. The "gunners" responsible for preparing and firing the howitzers are the first on the mountain. During their

Figure 7.10. Liam Fitzgerald stands beneath a Gazex avalanche-mitigation system on Mount Superior. Courtesy Adam Naisbitt.

Figure 7.11. Utah Transit Authority (UTA) bus hit by an avalanche in Little Cottonwood Canyon on December 23, 1988. Courtesy National Weather Service.

Former *Powder Magazine* editor Steve Casimiro once asked, "Did you ever wonder what would happen if it started snowing and never stopped?" The answer in Little Cottonwood Canyon is interlodge, a period during which you are confined to a public or private structure or building for protection while the avalanche hazard is high or avalanche-mitigation activities are under way. During interlodge, travel is not permitted between buildings, even by foot (figure 7.12). Within the town of Alta, violation of interlodge is a class B misdemeanor subject to possible imprisonment for up to six months and a fine of up to $1,000. During maximum-security interlodge, not only is travel restricted but people must go to the rooms or areas within houses or lodges that are most protected from avalanches. For some lodge guests and employees, this can mean being relocated to a basement or safe room for an extended period. This is similar to the action that might be taken during a tornado warning, but it is legally enforced and can last for hours or even days. In extreme circumstances, interlodge can last for more than a day, including fifty-two hours from February 6–8, 2020, and sixty hours from February 14–18, 2021.

Interlodge has a mythical status among skiers. The dream is to be at Alta or Snowbird on a day when interlodge ends and the ski areas open—but the Little Cottonwood highway remains closed. Locals call this a country club day because there's exclusive access to deep powder while thousands of Salt Lakers remain barred from entry at the bottom of the canyon. On the other hand, interlodge can torture skiers as they remain sequestered for hours or days while powder piles up outside. This can happen during multiday storm cycles when a sequence of winter storms in rapid succession keeps the avalanche hazard high.

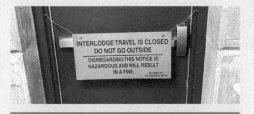

Figure 7.12. Thou shalt not pass! Stuck in powder purgatory during interlodge. Courtesy Joey Camps.

barrage, all nonessential personnel are confined to homes or lodges in the village of Snowbird or town of Alta. This period of restricted travel is known as **interlodge**. Then, additional snow-safety personnel perform avalanche-mitigation work at the ski areas using two-pound explosive charges thrown by hand (known as hand charges) or detonated while hanging from a wire over the snowpack. Increasingly, Wyssen avalanche towers and other avalanche-mitigation systems are being used. These efforts seek to disrupt weak layers and remove snow that could potentially avalanche. The people in this business are tough and frequently conduct these operations in extremely cold, hostile weather.

The snow conditions, weather, and results obtained during mitigation work dictate the opening and closing of terrain. Sometimes Mother Nature wins the battle, and some lifts and terrain are closed to skiers, or the ski areas may not open at all. This can be frustrating but think about the arsenal employed to reduce the avalanche hazard. When the patrol closes terrain, you don't want to be skiing there.

Using explosives and getting first tracks may be fun, but avalanche-mitigation work is dangerous. The American Avalanche Association recognizes seventy-three ski patrollers, highway technicians, snow rangers, avalanche forecasters, gunnery crewmembers, search-and-rescue personnel, guides, and other avalanche professionals who have been killed in the United States from the mid-1940s through the 2020–2021 season. Most died in avalanches, but others were killed in accidents involving explosives or artillery weaponry. Additionally, the American Avalanche Association also recognizes professionals who have died from suicide due to the mental challenges and traumatic stress experienced by those working in snow safety and search and rescue. All of this prompts the question: "Have you hugged a ski patroller today?"

CHALLENGES FOR THE FUTURE

Avalanche-mitigation work in Little Cottonwood Canyon is evolving. In recent years, the installation of Wyssen avalanche towers and Gazex avalanche-mitigation systems enables less reliance on 105-millimeter howitzers and reduces the risks associated with using military-grade weaponry. It is likely that more Wyssen avalanche towers and other avalanche-mitigation systems will be installed in the coming years, and use of the howitzers will be greatly reduced or eliminated altogether.

But the greatest challenge involves improving safety and traffic flows along the Little Cottonwood Canyon Road. Everyone agrees that the avalanche hazard along the road is unacceptably high and that closures have substantial consequences for businesses in the canyon, the quality of life for residents near the canyon mouth, and the skier experience.

During the summer of 2022, UDOT announced their preferred alternative for improving transportation in Little Cottonwood Canyon. It involves building a gondola from near the mouth of Little Cottonwood Canyon to Alta with a stop at Snowbird. Skiers would access the gondola base station via bus from mobility hubs near the mouth of Big Cottonwood Canyon and along 9400 South in Sandy. Snow sheds would also be added to the highway for avalanche protection along with parking lots at trailheads and infrastructure to toll vehicles driving up the canyon. The capital costs for implementing these changes are $592 million, with $7.6 and $3 million in winter and summer operations and maintenance costs annually.

The gondola is controversial, however, and it could take years to secure the necessary funding for its construction. The stakeholders in Little Cottonwood Canyon include the ski areas and their customers, communities at the base of the canyon, backcountry skiers and snowshoers, hikers and mountain bikers, climbers, tourists and visitors, and Utah residents who rely on the canyon for drinking water. Can the resources and political will be found to improve safety and access in a way that yields a net environmental benefit and safety? That is the challenge facing Little Cottonwood Canyon.

8 Beyond the Ropes

SKI AREA
BOUNDARY

NO SKI PATROL
OR AVALANCHE
CONTROL BEYOND
THIS POINT!

Greg and Loren woke up on an early January morning, dropped their three-year-old son off at day care, and drove to the Canyons Village base area at Park City Mountain Resort. Expert skiers, they probably took a few laps at the ski area but were lured by the untracked powder in the easily accessed back-country surrounding the Ninety-Nine 90 high-speed quad. They spoke with ski patrollers, who warned of high avalanche danger, but ultimately the desire to ski deep powder proved too great. They passed through a backcountry access gate and booted up a ridge to Square Top, a 9,800-foot peak with northeast-facing avalanche-prone slopes of more than thirty-five degrees.

That afternoon Greg and Loren did not pick up their son at day care. Friends and authorities began a search that evening. Where were they? Did they go skiing at Snowbird? Perhaps Alta? Eventually, a friend found their car in one of the parking lots at Canyons.

The next morning, someone spotted a boot track up Square Top and two short ski tracks from the summit that ended at a fracture line where a massive **slab avalanche** released from the slope. A search party of forty-five volunteers and ski patrollers located a ski and then, using long poles to probe the snow, two bodies beneath 1.5 to 3.5 feet of snow. Greg's hand was close to the surface.

US AVALANCHE FATALITIES

An average of twenty-five people die each year in avalanches in the United States.[1] More than 90 percent of these deaths occur in the backcountry, which includes lift-accessible terrain outside of ski-area boundaries, such as where Greg and Loren tragically perished. Most of the victims are outdoor recreationists, nearly all of whom are experts at their sport, including skiers and snowboarders, snowmobilers, hikers and snowshoers, and mountaineers.

Among ski and snowboard victims, 75 percent are earn-your-turns types who access the backcountry without the assistance of a lift. Another 25 percent enter the backcountry after riding a lift, either through open backcountry avalanche gates or illegally through closed backcountry gates and roped area boundaries.

An average of one avalanche fatality occurs per year in terrain that has been opened by the ski patrol at US ski areas. This represents only 4 percent of all skier and snowboarder avalanche fatalities, even though there are over 50 million skier visits per year at US ski areas, about 8.5 million of which involve skiing on ungroomed slopes that are steep enough to avalanche. This puts the odds of being killed by an avalanche in open, inbounds terrain at about 1 per 10 million skier days. Although not quite zero, these odds are quite long.

Some skiers and snowboarders refer to backcountry that can be accessed from a ski lift as **sidecountry** (if turns can be made with no climbing) or **slackcountry** (if turns require some climbing, either by booting or using **climbing skins**).[2] Don't let these names suggest that you are skiing anything but the backcountry. Whether you call it backcountry, sidecountry, or slackcountry, the snowpack is not the same as it is in bounds, where it has been subjected to substantive avalanche-mitigation efforts. The snowpack in bounds might seem absolutely bomber (meaning stable and safe), but as soon as you cross that rope line, you enter a new world where the avalanche hazard can be much higher. Don't let a safe, euphoric day of in-bounds powder skiing seduce you into thinking otherwise and making a deadly trip into the backcountry.

UTAH AVALANCHE FATALITIES

Utah averages 2.7 avalanche fatalities a year.[3] Backcountry skiers and snowboarders represent 56 percent of the victims, followed by snowmobilers with 41 percent. Of the backcountry skier and snowboard victims, 40 percent entered the backcountry after riding a ski lift, which is higher than the national average. Utah skiers also comprise 21 percent of all US avalanche victims who entered the backcountry after riding a lift, even though Utah is home to only 9 percent of the ski areas in the western United States.

AVALANCHE ESSENTIALS

Avalanches are torrents of moving snow that may contain rocks, dirt, or ice and come in several varieties. The deadliest is the slab avalanche, in which an entire plate of snow fractures from the snowpack and breaks into blocks and pieces as it rushes down the mountain, typically at speeds of sixty to eighty miles per hour, far faster than you can ski. Don't assume you can outski one.

Slab avalanches are responsible for most skier and snowboarder fatalities, as the slab frequently fractures above and around the victim or victims after they have skied partway down the slope. This was the case with the avalanche that killed Greg and Loren. The fracture at the top of the slab is called the **crown**, and the surface on which the avalanche slides is the **bed surface** (figure 8.1). Crowns range in height from a couple of inches to tens of feet, but even a shallow slab avalanche can be deadly.

Slab avalanches occur in snowpacks with a strong, cohesive layer of snow (the slab) sitting on top of a weaker layer of snow (figure 8.2). When it comes

Figure 8.1. The bed and debris from a slab avalanche that released during avalanche-mitigation activities in Snowbird's Mineral Basin. Courtesy Adam Naisbitt.

to weak layers, size doesn't matter. In some avalanches, the weak layer is as thin as a fingernail. Experienced backcountry travelers are constantly on the lookout for situations in which slab avalanches may occur. They look for signs of recent avalanche activity, dig avalanche test pits with a shovel or their hand to locate slabs and weak layers, and look and listen for cracking and "whumpfing" sounds that occur as they move across the snow and the weak layer collapses.

Loose-snow avalanches start at a point and fan out down the hill. One often sees several grouped on a slope (figure 8.3). Sometimes called **sluffs**, they form in freshly fallen snow or in wet snow when it is warm, sunny, raining, or drizzling. Although responsible for fewer fatalities, loose-snow avalanches can carry a skier over rocks or cliffs in steep terrain or deeply bury a skier in topographic depressions or gullies known as **terrain traps** (figure 8.4).

Figure 8.2. Using a variety of methods, strong and weak layers can be identified in the snowpack. Here, Chris Covington (UDOT) has isolated a strong slab of snow that is sitting on top of a thumbnail-thin weak layer above Little Cottonwood Canyon. In situations like this, the weight of a skier on a steep slope can trigger a slab avalanche. Courtesy Adam Naisbitt.

Slab and loose-snow avalanches can be dry or wet. **Dry avalanches** occur when there is little or no liquid water in the snowpack, whereas **wet avalanches** occur when warm weather, sun, rain, or drizzle cause water to move through and weaken the snowpack (figure 8.5). Some avalanches have a mixture of dry and wet characteristics.

Avalanches may be natural or triggered by humans. Natural avalanches occur when falling snow, wind-blown snow, or rain stress the snowpack and push it over the edge or when the snowpack is weakened by, for example, a spring thaw. In a human-triggered slide, a person or persons provides the stress needed to push the snowpack over the edge. Human-triggered slides may be intentional, such as is done during avalanche-mitigation operations, or inadvertent, as is the case in most avalanche accidents.

Figure 8.3. Loose-snow avalanches during wind loading on Alta's Mount Baldy. Courtesy Tyler Cruickshank.

Figure 8.4. A small gully like this is a classic terrain trap and should be entered only during stable conditions since an avalanche from the sidewalls could deeply bury a skier. Courtesy Tyler Cruickshank.

Figure 8.5. Liam Fitzgerald surveys the debris from a powerful wet slab avalanche in Hellgate Back Bowl above Little Cottonwood Canyon. Note the gouging of the bed surface and the depth and size of the debris pile. Avalanches of this type are nearly impossible to survive. Courtesy Adam Naisbitt.

AVALANCHE TERRAIN

Most (but not all) fatal avalanches start on slopes of thirty to forty-five degrees (figure 8.6). Black-diamond trails at ski areas are typically thirty to thirty-five degrees and double-black-diamond trails are steeper than thirty-five degrees, which means the best slopes for steep-and-deep powder skiing are also the slopes most prone to avalanche.

Although slopes outside thirty to forty-five degrees are less likely to avalanche, don't assume they are always safe. Slopes that are less than thirty degrees can generate avalanches in very unstable conditions, or they can be in the runout zone for avalanches beginning in steeper terrain farther up the mountain. Slopes steeper than forty-five degrees are dangerous enough without avalanches but add a small sluff and you could be carried over rocks and cliffs to a dire fate.

On some slopes, trees, bushes, and rocks can serve as **anchors** that help hold the snowpack in place, but there's a catch-22. If the anchors don't hold, those caught on the slope in an avalanche will take a beating and potentially suffer severe trauma as they are strained through the trees, bushes, and rocks. Many avalanche victims experience severe trauma, which in some cases either contributes to or causes their death.

Avalanche conditions typically vary depending on the direction a slope faces, which is known as **aspect**. Aspect affects the timing, intensity, and daily amount of energy received from the sun; the removal, transport, or deposition of snow by the wind; and, in some instances, cloud cover and related impacts on the snowpack. For example, in the Northern Hemisphere winter, south-facing aspects are more prone to melt-freeze cycles than north-facing ones. Wind can scour snow from the windward (upwind) side of a ridge and deposit snow on the lee (downwind) side of the ridge up to ten times faster than snow falling from the sky, greatly stressing the snowpack and forming **wind slabs** that are frequently unstable. Sometimes moving from one side of a ridge to the other can have grave consequences.

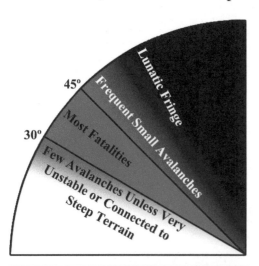

Figure 8.6. Most avalanche fatalities happen on slopes between thirty and forty-five degrees.

130

AVALANCHE RATINGS AND ASSESSMENTS

Avalanche hazard fluctuates greatly because of changes in the weather and snowpack. The avalanche danger can vary slowly, changing little from day to day, or rapidly. During heavy snow and high winds, or strong heating by the sun, the avalanche danger can increase in a matter of minutes.

For winter recreationists, the Utah Avalanche Center (utahavalanchecenter. org) issues avalanche advisories for several backcountry areas in Utah, including the Wasatch Mountains near Salt Lake City, Ogden, and Provo (figure 8.7).[4] These advisories are based on snowpack observations, reports from other avalanche professionals and backcountry enthusiasts, and an assessment of the past, current, and future weather. Danger ratings follow the **North American Public Avalanche Danger Scale**, which has five levels: low, moderate, considerable, high, and extreme (figure 8.8). For example, when the avalanche danger is moderate, natural avalanches are unlikely, but human-triggered avalanches are possible. Although there are times when a single rating applies, the danger typically varies with elevation and aspect, as summarized with an **avalanche danger rose** (see figure 8.7). A written threat assessment discusses specific concerns. Even with all this information, the avalanche advisory only provides guidance on the likelihood, size, and mechanics of potential ava-

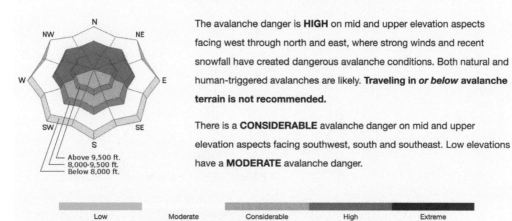

Forecast for the Salt Lake Area Mountains
Issued by Trent Meisenheimer for Tuesday, December 28, 2021

Above 9,500 ft.
8,000-9,500 ft.
Below 8,000 ft.

The avalanche danger is **HIGH** on mid and upper elevation aspects facing west through north and east, where strong winds and recent snowfall have created dangerous avalanche conditions. Both natural and human-triggered avalanches are likely. **Traveling in *or below* avalanche terrain is not recommended.**

There is a **CONSIDERABLE** avalanche danger on mid and upper elevation aspects facing southwest, south and southeast. Low elevations have a **MODERATE** avalanche danger.

Low   Moderate   Considerable   High   Extreme

Figure 8.7. Avalanche advisory summary statement issued by the Utah Avalanche Center including an avalanche rose (left) and summary narrative (right). Courtesy Utah Avalanche Center.

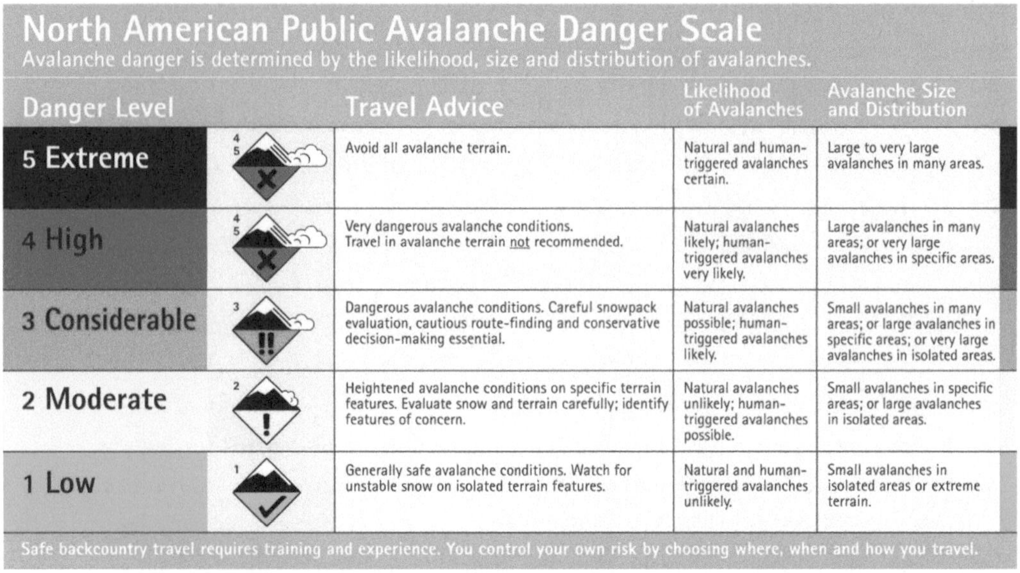

North American Public Avalanche Danger Scale
Avalanche danger is determined by the likelihood, size and distribution of avalanches.

Danger Level		Travel Advice	Likelihood of Avalanches	Avalanche Size and Distribution
5 Extreme		Avoid all avalanche terrain.	Natural and human-triggered avalanches certain.	Large to very large avalanches in many areas.
4 High		Very dangerous avalanche conditions. Travel in avalanche terrain not recommended.	Natural avalanches likely; human-triggered avalanches very likely.	Large avalanches in many areas; or very large avalanches in specific areas.
3 Considerable		Dangerous avalanche conditions. Careful snowpack evaluation, cautious route-finding and conservative decision-making essential.	Natural avalanches possible; human-triggered avalanches likely.	Small avalanches in many areas; or large avalanches in specific areas; or very large avalanches in isolated areas.
2 Moderate		Heightened avalanche conditions on specific terrain features. Evaluate snow and terrain carefully; identify features of concern.	Natural avalanches unlikely; human-triggered avalanches possible.	Small avalanches in specific areas; or large avalanches in isolated areas.
1 Low		Generally safe avalanche conditions. Watch for unstable snow on isolated terrain features.	Natural and human-triggered avalanches unlikely.	Small avalanches in isolated areas or extreme terrain.

Safe backcountry travel requires training and experience. You control your own risk by choosing where, when and how you travel.

Figure 8.8. The North American Public Avalanche Danger Scale. Courtesy American Avalanche Association and Canadian Avalanche Association.

lanches. Ultimately, backcountry travelers must perform their own assessment and make their own decisions about where to travel.

Curiously, many avalanche fatalities don't occur on the days with the highest avalanche danger. In the United States, 38 percent of avalanche fatalities occur on high hazard days and only 2 percent occur on extreme hazard days (figure 8.9). The majority (58 percent total) occur on moderate and considerable hazard days. This is because the terrain, snowpack, and weather aren't the only players in this potentially deadly poker game.

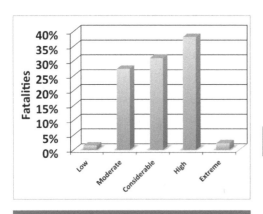

Figure 8.9. Percentage of US avalanche fatalities by danger rating. Source: Greene et al. (2006).

THE HUMAN ELEMENT

At the heart of nearly all avalanche accidents lies the inability of humans to recognize the existence or severity of the avalanche hazard. Fatal avalanches don't just sneak up on most victims out of the blue. *In over 90 percent of all avalanche deaths, the victim or some-*

one in the victim's party triggers the avalanche. As the Northwest Avalanche Center website asserts, "Choice not chance causes most avalanche accidents."

When the avalanche danger level is high or extreme, most people elect not to venture into avalanche terrain—thus there are fewer fatalities, and many of those are due to ignorance or lack of knowledge about the severity of the danger. When the avalanche danger is low, most terrain can be safely skied or snowboarded, and it is difficult (but not necessarily impossible) to trigger a dangerous avalanche. When the danger is moderate or considerable, people tend to be more confident venturing out in search of powder. Safe travel and skiing are possible on some terrain, but perhaps not all. Accidents happen when the perception of risk does not match reality.

If we were all Vulcans or droids who thought logically, our perception of risk would stand a better chance of matching reality. In *The Empire Strikes Back*, C-3PO nervously tells Han Solo, "Sir, the possibility of successfully navigating an asteroid field is approximately 3,720 to 1." Han Solo responds forcefully, "Never tell me the odds." Inspired by confidence and driven by the fear of being caught by Imperial Forces, he elects to enter the asteroid field anyway.

Being a movie hero, Han Solo is all about beating the odds, but in the backcountry, Mother Nature does not discriminate. She doesn't care if you are an intermediate or expert skier or snowboarder, or if you are a backcountry greenhorn or a snow-safety professional. Accidents happen in the backcountry either because we don't recognize the risk or because we consciously (like Han Solo) or unconsciously chose to ignore the warning signs.

POWDER FEVER AND OTHER PITFALLS

Avalanche victims include not only skiers and snowboarders who enter the backcountry from ski areas but also seasoned "earn-your-turns" skiers and snowboarders with many years of backcountry travel experience. As a friend once told me on a one-week backcountry trip in British Columbia, "Even the guides aren't infallible." Among victims with some avalanche education, human factors are responsible for 82 percent of the accidents, with lack of judgment, rather than lack of knowledge, accounting for most incidents. We are all human and prone to errors in judgment.

In particular, we all "suffer" from **powder fever**. Instead of assessing the avalanche hazard objectively, our reality is clouded by our desire to have that epic powder run. This biases the way we gather and interpret evidence. We see people going through a backcountry avalanche gate and instead of questioning if they are aware of the hazard and have done an assessment, we assume they must know something we don't and that the hazard may not be as high as advertised. We see them take a run and assume the entire backcountry is safe, rather than considering the steepness and aspect of the slope they skied or the possibility that they simply didn't hit the "sweet spot" that will release an avalanche.

We are also influenced by groupthink and peer pressure. You see cracking and hear whumpfing, but all your buddies are convinced this is nothing to worry about, so you assume your concerns are unwarranted and don't speak up. You have been boasting for months about your big vacation to Utah, but the ski area is all tracked out and you want that bottomless powder run to boast about on Instagram. A few people have skied the backcountry, so you assume it must be safe.

Sometimes familiarity plays a role. When you ski a run many times without incident, it builds confidence. But the avalanche danger can vary dramatically and change quickly. The run that was bomber yesterday could be a death trap today.

Pitfalls like those described above represent human factors that contribute to avalanche accidents, which avalanche researcher Ian McCammon cleverly summarized as a single acronym, **FACETS**, with each letter corresponding to a human factor that biases our judgement. F is for Familiarity, which represents how our decisions are biased by past experiences (e.g., we have skied this run many times before and thought it was safe). A is for Acceptance, which is the tendency to do things to impress or be accepted by peers (e.g., we wanted to get a great powder shot to post on Instagram). C is for Commitment/Consistency, which is the tendency of individuals to stick to existing plans to meet timelines and other limitations (e.g., it was our last chance to ski this couloir this season). E is for Expert halo or the tendency to accept the assessments of others who have greater perceived expertise (e.g., Jonny has been backcountry skiing in the Wasatch for several years, and we were reluctant to question his assessment). T is for Tracks, which is an intense desire to ski a line first or get untracked powder before a slope is tracked up. In other words, powder fever. Finally, S is for Social facilitation, which is the influence that being in a group has on you (e.g., I had concerns about the avalanche danger, but everyone else thought it was fine, so I went with their decision).

These human factors are readily apparent in many backcountry avalanche accidents, and they don't work in isolation. Often, they have a cumulative effect. Consider your thought process after a long approach with four friends to the top of a run you've wanted to ski for years on a bottomless powder day. Your friends want to ski it and get that Instagram shot, but you have reservations. Many of the FACETS apply on such a day.

GOING BEYOND THE ROPES

This chapter has provided an overview of avalanches and the circumstances that contribute to accidents. It is not a guidebook for safe backcountry travel, although hopefully it illustrates the role that human factors play in most avalanche accidents.

If you are thinking of backcountry skiing or snowboarding, you might consider the following to get started. Read *Staying Alive in Avalanche Terrain* by Bruce Tremper and *The Avalanche Handbook* by David McClung and Peter Schaerer. Buy a new digital avalanche beacon and practice searches with your friends. Get a probe and a metal avalanche shovel and learn how to use them. Visit your local avalanche center website and take avalanche classes that cover avalanche awareness, snowpack assessment, decision making, hazard management, and avalanche rescue. Find good partners—those who have good heads on their shoulders, even if they may not be the strongest skiers. Finally, check the avalanche report, assess the snow conditions yourself, and adjust plans as conditions dictate. It is better to ski and snowboard conservatively and enjoy a lifetime of powder than to put it all on the line for one moment of bliss.

NOTES

1. Based on data from the Colorado Avalanche Information Center for 2011/2012–2020/2021.

2. Definitions from the late Jackson Hole skier Steve Romeo and his tetonat.com blog.

3. Based on data from the Utah Avalanche Center for 2011/2012–2020/2021.

4. See avalanche.org for general avalanche information and advisories for other regions of the country.

9 Powder Prediction

Prediction is very difficult, especially about the future. Those words, famously quipped in one form or another by everyone from Nobel Prize–winning physicist Niels Bohr to Hall of Fame baseball catcher Yogi Berra, are figuratively tattooed on the shoulder of every professional or armchair meteorologist who has tried to forecast for the Wasatch Mountains. Our storms are chewed into pieces by the Sierra Nevada, Cascade Mountains, and other upstream mountain ranges; our topography is super steep and narrow; and we have this salty puddle of water known as the Great Salt Lake that sometimes teases us with lake effect following the passage of a cold front. All of these things make powder prediction in the Wasatch Mountains very difficult. Nevertheless, weather observations and forecasts are getting better every day and can help you anticipate when to call in sick and where to find the deepest powder.

EASY-ACCESS WEATHER INFORMATION AND FORECASTS

For skiers and snowboarders who simply want reliable weather information and forecasts, I recommend the following websites. The first is utahavalanchecenter.org, which provides highly detailed avalanche and mountain weather advisories. These are prepared for backcountry travelers, but they are also extremely useful for getting a handle on weather at the ski areas, although the avalanche advisories and ratings apply only to backcountry areas, not within ski area boundaries where avalanche-mitigation work is done. Next is the website for the Salt Lake City National Weather Service Forecast Office (https:// www.weather.gov/slc/). In addition to weather observations, discussions, and forecasts, the National Weather Service also produces a detailed forecast for the Cottonwood Canyons (https://www.weather.gov/slc/mtnwx?Cottonwood). There's also opensnow.com, run by Joel Gratz and his team of hard-skiing meteorologists and weather enthusiasts.

Another great site is wasatchsnowinfo.com/Wasatch, which specializes in ski weather and is the brainchild of Chris Larson, a University of Utah computer science alum with a self-described "PhD in powder skiing at Alta." It provides point-and-click, one-stop shopping for snow reports, avalanche and snowpack information, observations from mountain weather stations, forecasts, and webcams.

PRODUCING YOUR OWN WEATHER FORECASTS

The sites above are great for basic information, but surely you don't want your powder pursuits to depend on the whims of professional "weather guessers."

Perhaps you can do better? I know some amateur weather forecasters who are quite good, largely because they have developed a great feel for the weather and climate where they like to ski or snowboard. They learned the basics of weather forecasting, started forecasting using products readily available on the Internet, and now have a leg up on the competition when it comes to getting the goods.

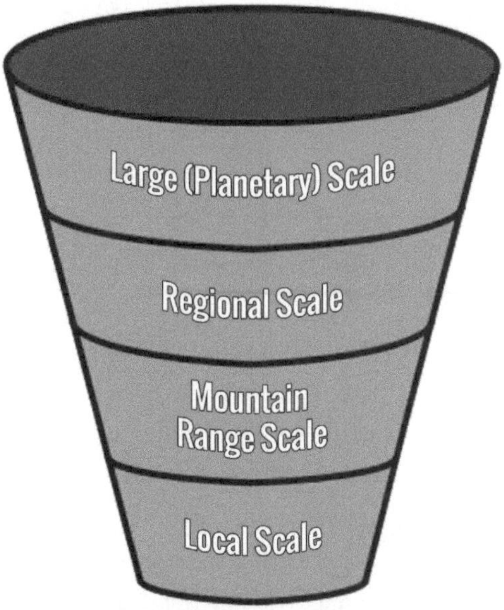

Meteorologists use a technique known as the **forecast funnel** to predict the weather in mountainous regions (figure 9.1). The forecast funnel starts with what meteorologists call the large scale. Begin by getting a handle on the location and movement of high- and low-pressure systems, fronts, and troughs and ridges at jet-stream level. Then, funnel down to smaller scales and evaluate how the interaction of these systems with the regional and local topography will influence the weather at the ski or backcountry area of interest.

Figure 9.1. The forecast funnel. Adapted from Snellman 1982; Horel et al. 1988.

For example, suppose there is a moisture-laden cold front moving across Utah today and you want to ski powder at two different ski areas over the next two days. The **computer models** suggest that ahead of the front the flow will be southwesterly, whereas behind the front it will be northwesterly. That's the big picture—now funnel down and add in the topographic effects. During southwesterly flow storms, ski areas like Sundance and Snowbasin usually (but not always) see strong orographic precipitation enhancement (see chapter 2). Either of these might be a good option for your first day. On the other hand, tomorrow, when the front has passed, you might want to hit a ski area in the Cottonwood Canyons, which are favored in northwesterly flow.

That example makes it sound easy, but usually weather forecasting is more complicated. For example, the moisture might be confined to near the front, which is sagging slowly into Utah from the northwest. This could mean that Snowbasin, which is farther north, will get more snow than Sundance. On the other hand, Mount Timpanogos is a huge mountain, about 2,000 feet higher than the mountains around Snowbasin. If the depth of the upslope flow is important, perhaps Sundance will get more of the goods. When it comes to detailed forecasts you can't rely on generalizations. You need to know how to use forecast tools like satellites, radar, mountain weather cameras, mountain weather stations, and computer models.

Satellites

Satellites provide a bird's-eye perspective on the atmosphere, usually from an altitude of about 22,000 miles, although there are some weather satellites that fly in orbits as low as a few hundred miles. The 22,000-mile altitude allows for a **geostationary orbit** that enables continuous monitoring at a fixed point above the Earth's surface. Lower satellite orbits enable higher resolution images and monitoring using specialized instruments such as space-borne radars (yes, we have those!), but the area observed is smaller and the speed of the satellite and rotation of the Earth do not allow for continuous monitoring over any one area. Most of the satellite imagery you've seen has probably been collected by US National Oceanic and Atmospheric Administration (NOAA) Geostationary Operational Environmental Satellites, known as GOES. The current GOES generation provides continuous monitoring of the western hemisphere in sixteen light-energy ranges known as bands. There are two visible bands, four near-infrared bands, and ten infrared bands.

Visible satellite images based on one or more visible bands provide the closest perspective to that of the human eye and are sensitive to how much sunlight is reflected by clouds, snow, ice, and the Earth's surface (figure 9.2a). Clouds are very reflective and can be readily identified in visible satellite images if they lie over a darker surface like the ocean or land that is not snow covered. When the land is snow covered, however, it acts as a sort of camouflage and makes cloud identification using visible imagery more difficult. Black-and-white visible images are often created using observations from a single visible band, but "true-color" visible images are created using observations from two visible bands and one near-infrared band and provide color imagery like what we might see with our own eyes.

Infrared (IR) satellite images provide the meteorological equivalent of seeing the world through night-vision glasses (figure 9.2b). Warmer objects emit more infrared radiation than colder objects. Because temperature usually decreases with height, we can usually distinguish clouds with cold, high cloud tops from clouds with warm, low cloud tops. Most images that are commonly referred to as infrared are based on either a "clean" or "window" IR band that is not strongly affected by atmospheric water vapor, enabling

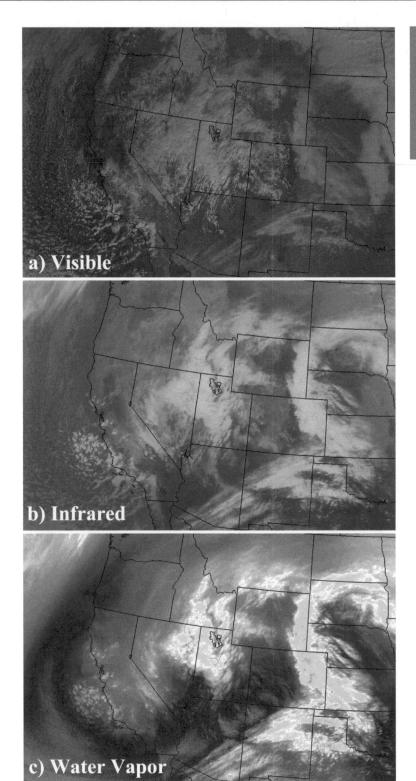

Figure 9.2. (a) Visible, (b) infrared, and (c) water vapor satellite images as a storm approaches Utah at 11:00 a.m. on February 8, 2013.

a) Visible

b) Infrared

c) Water Vapor

images that "see" through clear areas and better illustrate clouds or the Earth's surface.

Water vapor images are based on a special type of infrared radiation that is affected not only by clouds but also by water vapor. There are three water vapor bands measuring different layers of the atmosphere. Upper-level water vapor images are sensitive to water vapor near jet-stream level (20,000–35,000 feet [figure 9.2c]). This allows meteorologists to track water-vapor features (that is, patches of dry and moist air) to determine the strength and direction of the flow at jet-stream level even in areas that are cloud free. Low- and mid-level water vapor images are used to track winds and monitor storm environments in other atmospheric layers.

Visible, infrared, and water-vapor images are important tools for weather forecasting. However, these types of satellite images mainly tell us what is happening at the top of the clouds rather than in the clouds. Infrared satellite imagery, for example, can be very challenging to interpret. Most of the color or gray shades used for infrared satellite images emphasize the coldest cloud tops. While this is useful for identifying high clouds, it is not necessarily useful for determining where it is snowing. Over Utah, for example, many high clouds that look cold and menacing in infrared satellite images are actually quite thin and don't produce any snow. On the other hand, low clouds with relatively warm cloud tops produce many Wasatch Mountain snowstorms. Low clouds can be difficult to identify in infrared satellite images and, if there is snow on the ground, even visible satellite images. The bottom line is that it is best to consult multiple satellite bands and use them in conjunction with other tools, especially radar.

Radar

Radar provides a meteorological CAT scan of winter storms. The National Weather Service radar on Promontory Point, a mountain peninsula of the Great Salt Lake, provides the best information about Wasatch Mountain winter storms and is identified by the call letters KMTX (figure 9.3). KMTX sends out pulses of radio waves as it completes a series of circular scans at increasing tilt angles to the horizon. A small amount of this energy scatters off particles in the atmosphere, including raindrops and snowflakes, and returns to the radar. The intensity of this returned energy is measured by the radar receiver and is known as the **radar reflectivity**.

Images of radar reflectivity help us determine if and how hard it is snowing or raining and enable us to examine the evolution of precipitation features

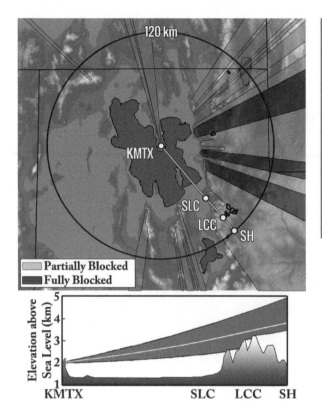

Figure 9.3. Coverage and scanning characteristics of the KMTX radar. Top: Areas where the lowest radar scan is partially or fully blocked by topography. Bottom: Height and width of the lowest radar scan over Salt Lake City (SLC), Little Cottonwood Canyon (LCC), and Soldier Hollow (SH), including partial blockage by the high terrain around the Cottonwood Canyons. White line depicts the center of the beam.

produced by weather systems, the Great Salt Lake, and the Wasatch Mountains (figure 9.4). The snowfall or rainfall rate is usually greatest where the radar reflectivity is the highest, although there are some important exceptions to this rule. We can also make a short-range forecast by tracking the movement and intensity of precipitation features, which I do frequently on my smart phone when skiing.

Radar imagery is strongly affected by the tilt of the radar, curvature of the Earth, atmospheric stability, and topography (figure 9.3). The 0.5-degree tilt of the KMTX radar is frequently used to examine storms in the Wasatch Mountains. The center of the radar beam for this tilt's circular scan is positioned at a 0.5-degree angle relative to the local horizon. This slight tilt, combined with the curvature of the Earth, means that the altitude of the beam relative to the Earth's surface increases with distance from the radar. This increase in altitude is partly offset by atmospheric stability effects that cause the beam to bend back slightly toward the Earth. Although the amount of this bending, known as **refraction**, varies, on average the center of the beam is about 2 kilometers (6,500 feet) above sea level (2,500 feet above the Great Salt

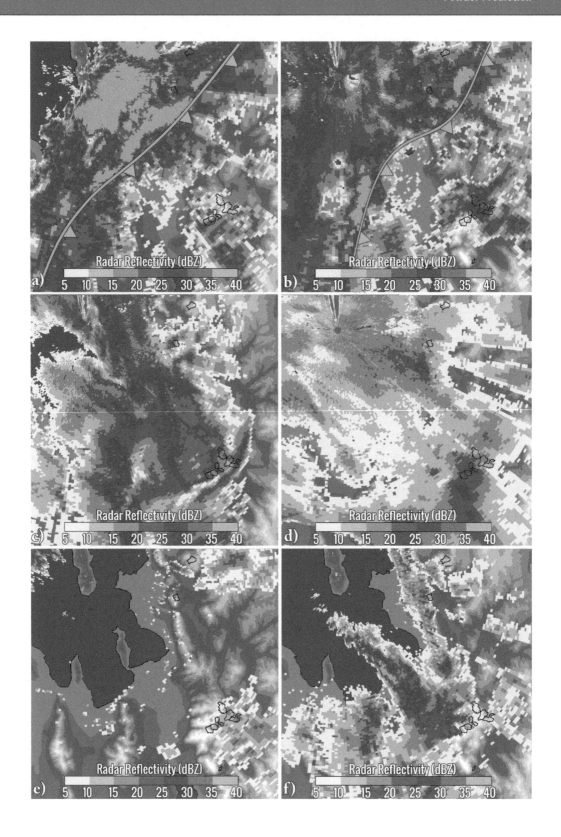

a) Radar Reflectivity (dBZ) 5 10 15 20 25 30 35 40

b) Radar Reflectivity (dBZ) 5 10 15 20 25 30 35 40

c) Radar Reflectivity (dBZ) 5 10 15 20 25 30 35 40

d) Radar Reflectivity (dBZ) 5 10 15 20 25 30 35 40

e) Radar Reflectivity (dBZ) 5 10 15 20 25 30 35 40

f) Radar Reflectivity (dBZ) 5 10 15 20 25 30 35 40

Figure 9.4 (previous page). Examples of precipitation features observed by the KMTX radar. (a) Cold front at 8:37 p.m. on December 2, 2012. (b) Cold front at 7:22 p.m. on February 16, 2011. (c) Precipitation enhancement over the eastern Salt Lake Valley and lower Cottonwood Canyons at 6:26 a.m. on April 6, 2012. (d) Precipitation enhancement over Mount Timpanogos and the upper Cottonwood Canyons at 7:28 a.m. on December 18, 2012. (e) Precipitation enhancement over the Wasatch Back at 4:54 p.m. on February 25, 2011. (f) A lake-effect snow band with orographic precipitation over the Wasatch Mountains at 5:47 a.m. on November 5, 2011.

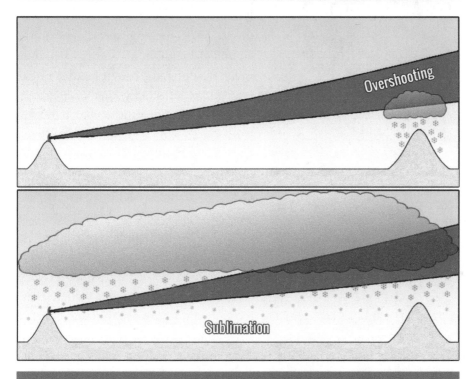

Figure 9.5. Schematic depictions of radar overshooting a shallow mountain storm (top) and radar overshooting sublimation within dry air over a valley (bottom).

Lake) at the radar, 2.75 kilometers (9,000 feet) above sea level over Salt Lake City (4,700 feet above the valley floor), and 3 kilometers (10,000 feet) above sea level over Little Cottonwood Canyon. The radar beam also spreads with distance from the radar.

Knowing the altitude of the radar beam is important for radar interpretation. Some mountain snowstorms are very shallow and may be overshot or only partially sampled at longer ranges (figure 9.5). On the other hand, because the center of the radar beam is several thousand feet over valleys, including

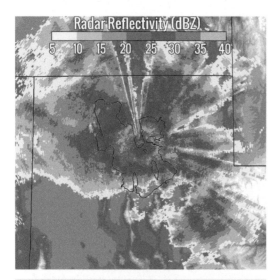

Figure 9.6. Example of a bright band encircling the KMTX radar location at 2:31 p.m. on December 19, 2010.

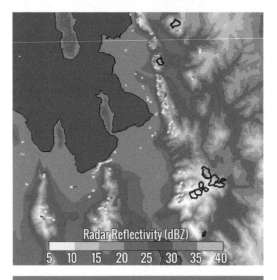

Figure 9.7. Example of ground clutter at 9:26 a.m. on December 21, 2012.

the Salt Lake Valley, there are times when the snowflakes or raindrops it intercepts sublimate or evaporate before they reach the valley floor. In this instance, the radar reflectivity suggests it is snowing or raining harder than it really is over lowland locations. Finally, the radar beam sometimes cuts through the layer where snowflakes are melting and becoming raindrops. Wet, melting snowflakes scatter an unusual amount of energy back to the radar, resulting in a **bright band**, a ring of high radar reflectivity that encircles the radar site during storms with high freezing levels (figure 9.6). The bottom line is that it is usually a good idea to consult observations from precipitation gauges for a sanity check when using radar imagery.

Mountains create additional problems for radar interpretation. Sometimes they "block" the radar beam (figure 9.3). When this happens, a huge amount of energy, known as **ground clutter**, is scattered back to the radar (figure 9.7). Ground clutter is sometimes automatically removed from radar images, leaving an empty area in which there is no information. If the beam is fully blocked, no information can be obtained from the other side of the mountains. If the beam is partially blocked, some of the radar signal can return, but at an intensity that is lower than it would be in the absence of blocking. Therefore, some areas over and to the east of the Wasatch Mountains have no radar coverage or a bias toward lower radar reflectivity values.

Finally, energy from the radar can be scattered by more than just precipitation. For example, it is not uncommon to see migrating birds or **chaff**, a

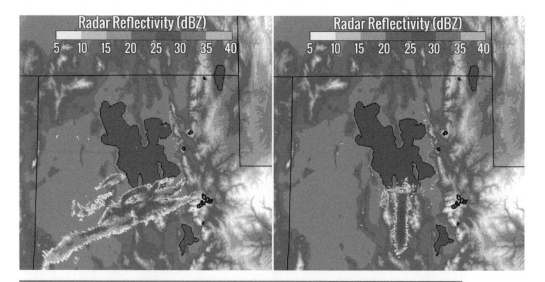

Figure 9.8. Examples of nonmeteorological radar features produced by chaff at 8:01 p.m. on April 26, 2006 (left) and migrating birds (most likely eared grebes) at 6:08 p.m. on December 24, 2009 (right).

military countermeasure used to confuse radars (and frequently used for military training in the western United States [figure 9.8]). High radar reflectivities produced by birds and chaff are usually fairly obvious but can look like snow bands. Don't be fooled!

Mountain Weather Cameras

Mountain webcams are the ultimate eye candy for meteorologists. On big storm days when I'm stuck in the office, I simply cannot take my eyes off them. Given some of the shortcomings of satellite and radar images, the webcam is the next best thing to being there.

Every major ski area in Utah has one or more webcams. Snag an image on a clear day for comparison with storm days and learn their locations and orientations so you can extract as much information as possible. For the ultimate stimulation, you can animate images from snowcams like Snowbird's (figure 9.9).

Mountain Weather Stations

Weather cams are great, but they don't tell you how cold it is, how hard the wind is blowing, or, with exceptions like the Snowbird snowcam, how much

Figure 9.9. Seeing is believing. The Snowbird snow cam shows seven inches of snow overnight. Courtesy Snowbird.

snow has fallen. Fortunately, most of the ski areas in Utah provide access to observations from their mountain weather stations. The coverage in and around the Cottonwood Canyons and Park City ski areas is incredible, with data from more than forty stations (figure 9.10). Most provide only wind and temperature observations, but a few provide automated snow measurements.

It pays to know precisely where these stations are as mountain weather changes dramatically over short distances. For avalanche-mitigation purposes, some are located on highly exposed ridges that tend to be windier, sometimes much windier, than the ski area. For example, the Mount Baldy observing site at Alta is typically far windier than the more sheltered ski terrain below (figure 9.11).

In addition, most of the temperature sensors on these stations are unaspirated, which means they don't have a fan that blows air through them. The aspiration is done by the wind, which is fine except when the air is calm. On a sunny day with light winds, the sensor overheats, yielding a temperature that is higher than that of the ambient air. A reading of 38°F on a clear, calm afternoon in March might not necessarily mean the actual temperature was above freezing. Perhaps the snow will not be topped by a crust in the morning.

On most weather maps, a schematic diagram known as a station model illustrates the weather at each station (figure 9.12). A line extending from the station

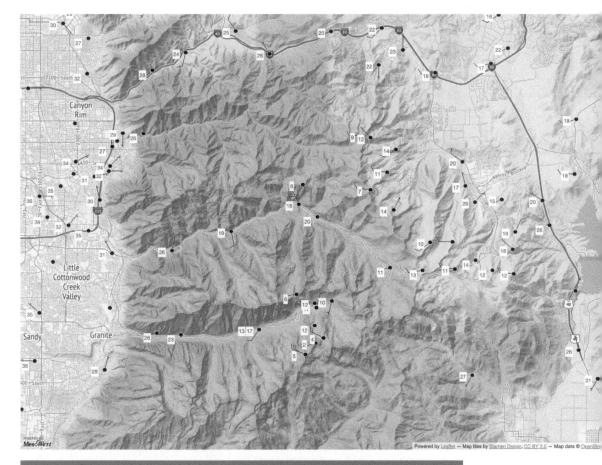

Figure 9.10. Surface weather observations in the in the region surrounding the Cottonwood Canyons and Park City ski areas at 2:15 p.m. on April 12, 2022. Courtesy Mesowest (mesowest.utah.edu). Powered by Leaflet—Map tiles by Stamen Design, CC BY 3.0—Map data © OpenStreetMap.

location indicates the direction from which the wind is blowing. Half barbs (five knots),[1] full barbs (ten knots), or flags (fifty knots) attached to this line can be added up to determine the wind speed. Temperature and dewpoint are usually plotted to the upper left and lower left, respectively. Most automated mountain weather stations do not collect visibility, pressure, or present weather observations (snow, rain, and so on), but this information can be included if available.

Although I like to look at weather maps, I also spend a good deal of time examining **meteograms**, graphs that contain a time series of meteorological observations (figure 9.13). I like short (one- or two-day) time series for examining weather details but also longer time series (sometimes as long as a month)

Figure 9.11. Observation locations in the Collins Gulch area of Alta ski area. Courtesy Peter Veals.

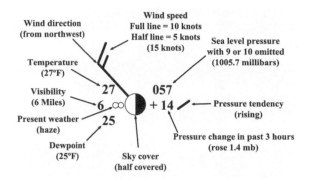

Figure 9.12. Common elements of a meteorological station model.

to get a handle on the recent climate at a site. Longer time series can also be useful for understanding snowpack evolution for avalanche forecasting.

There are a few weather stations that I watch religiously, but my favorite is the Alta-Collins observing site, which is located at 9,662 feet at Alta ski area (just above mid-mountain) and is identified by the call letters CLN (figure 9.11). CLN does not have a wind sensor, but it has the best upper-elevation precipitation gauge and snow-depth sensor in Utah. It's great to examine during storms.

SNOTEL (SNOwpack TELemetry) stations operated by the Natural Resources Conservation Service at a few ski areas (Park City, Snowbird, and Brighton) and throughout the Wasatch backcountry are also very useful. SNOTEL stations

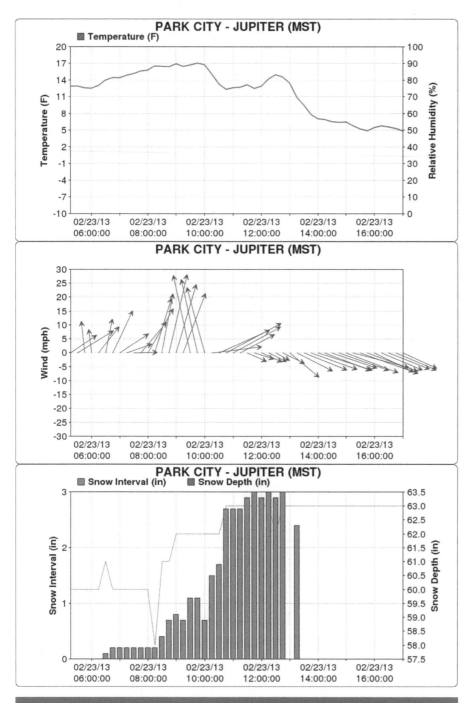

Figure 9.13. Meteograms of temperature (top), vector wind (middle), and recent and total snow depth (bottom) at the Park City Jupiter Bowl observing site. During this period, a cold-frontal passage occurred just after 10:00 a.m., producing about three inches of new snow. Courtesy Mesowest (mesowest.utah.edu).

Figure 9.14. A SNOTEL station. Courtesy USDA / NRCS.

measure the water content of the snowpack using an automated device known as a snow pillow that weighs the snow. They also measure precipitation (snow water equivalent) and snow depth (figure 9.14). Hourly SNOTEL snowpack and precipitation observations are available but can sometimes be difficult to interpret. The daily (twenty-four-hour) observations, however, are very useful, especially during big storms. I often use the SNOTEL snowpack snow-water equivalent observations for backcountry trips since they provide information on how robust the snowpack is in remote areas.

Computer Models

Computer models provide the backbone for weather forecasting and represent one of the great scientific achievements of modern times. Producing a forecast on a computer is sometimes called **numerical weather prediction** (NWP for short) and involves solving a complex set of mathematical equations that describe the processes that affect the weather. Most of these equations are based on physical laws, including the conservation of energy, conservation of mass, and conservation of momentum.

Meteorologists first dreamed of using numerical weather prediction to forecast the weather in the early twentieth century, but the dream wasn't realized until the development of ENIAC, the world's first general-purpose electronic computer, in the early 1950s. Since then, meteorologists have developed increasingly sophisticated and detailed computer models that take advantage of rapidly advancing computing, networking, and observational capabilities. Some of the largest computers in the world are dedicated to weather forecasting. In 2022, the Atos BullSequana XH2000 supercomputer at the European Center for Medium-Range Weather Forecasts (ECMWF), which produces the most accurate medium-range weather forecasts in the world, consisted of four compute clusters that collectively contained over one million processors (most personal computers have one to four) and performed 30 quadrillion (30,000,000,000,000,000) calculations per second.

Many computer models divide the atmosphere into a series of vertically layered **grid cells** (figure 9.15). The length of the side of a grid cell is the **grid spacing**. The smaller the grid spacing, the more resolved and detailed the model terrain and forecast. Higher-resolution models, however, also require more calculations (and hence computer power), and this is what ultimately places a limit on model resolution.

As of 2022, the National Centers for Environmental Prediction, a division of the National Weather Service, runs several operational computer models, including the **Global Forecast System (GFS),** the **North American Mesoscale Forecast System (NAM)**, and the **High-Resolution Rapid Refresh (HRRR)**. The GFS has a grid spacing of about thirteen kilometers and generates forecasts covering the entire world four times a day out to sixteen days in the future.

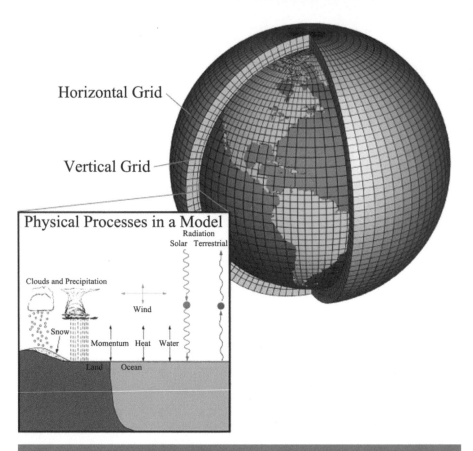

Horizontal Grid

Vertical Grid

Physical Processes in a Model

Radiation
Solar   Terrestrial

Clouds and Precipitation

Wind

Snow

Momentum   Heat   Water

Land   Ocean

Figure 9.15. Schematic diagram of a grid cell model. Adapted from a figure from the National Oceanic and Atmospheric Administration.

The NAM has a grid spacing of about twelve kilometers and produces forecasts for North America four times a day out to eighty-four hours (there is also NAM "nest" that covers the continental United States with three-kilometer grid spacing, although the skill of this nest is low for winter storms in the western United States). The HRRR has a grid spacing of about three kilometers. For the continental United States, it produces forecasts every hour out to eighteen hours, with an extension to forty-eight hours every six hours.

When using forecasts from these computer models, it is important to know how well they resolve the Wasatch Mountains. Models that run at grid spacings larger than five kilometers, such as the GFS, poorly resolve the Wasatch Mountains (figure 9.16). This is why the forecast funnel is so important. Use the forecasts from these models to understand the big picture, then use experience and knowledge of effects to infer how the Great Salt Lake and Wasatch

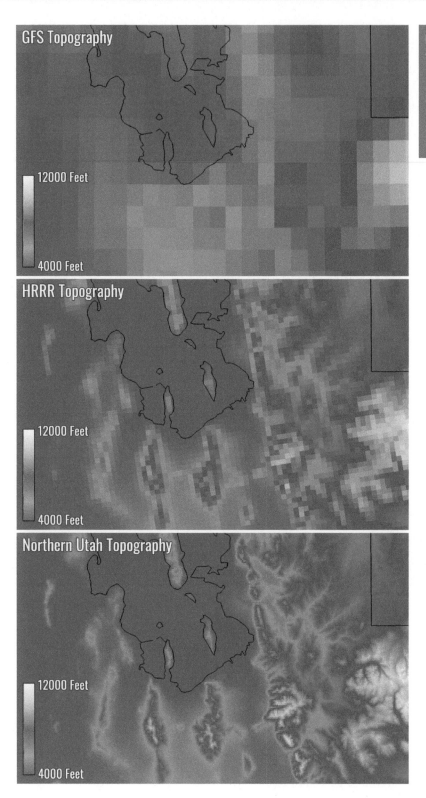

GFS Topography

12000 Feet

4000 Feet

HRRR Topography

12000 Feet

4000 Feet

Northern Utah Topography

12000 Feet

4000 Feet

Figure 9.16.
Topography of GFS model (top), HRRR model (middle), and northern Utah (bottom).

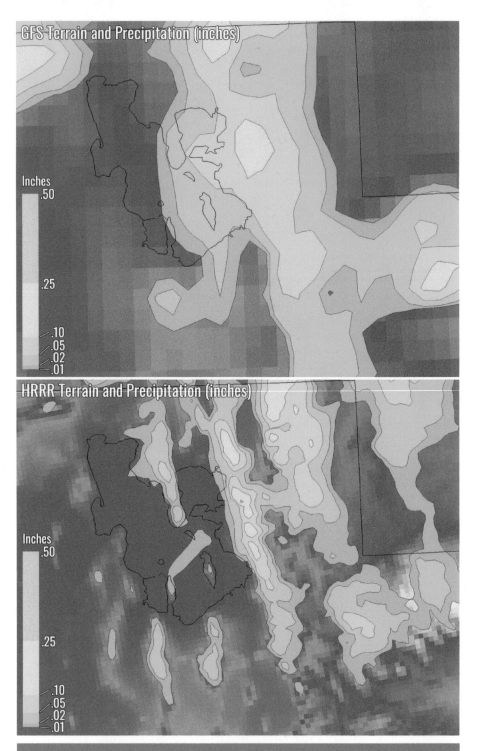

Figure 9.17. GFS (top) and HRRR (bottom) three-hour accumulated snow-water equivalent (in inches) forecast valid at 2:00 p.m. on March 8, 2021.

Mountains will further influence the weather. The HRRR has higher resolution, but still fails to resolve the fine-scale characteristics of the corrugated topography of the Cottonwood Canyons and surrounding ridgelines. Nevertheless, the three-kilometer grid spacing of the HRRR still better resolves the terrain and its influence on precipitation than the thirteen-kilometer grid spacing of the GFS, resulting in more detailed precipitation forecasts (figure 9.17).

Higher-resolution forecasts, including those generated with the **Weather Research and Forecast (WRF) model** (pronounced "Worf," like the Klingon in *Star Trek: The Next Generation*), are sometimes produced by university or private-sector groups and can better resolve fine-scale topography. Higher resolution, however, is not always the holy grail of weather forecasting. High-resolution model forecasts produce many **false alarms**, forecasts of major snowstorms that never materialize. These false alarms are related to small errors in the position, movement, and structure of large-scale weather systems like cyclones and fronts. For example, a thirty-six-hour WRF forecast might call for moist southwesterly flow to sag into northern Utah and produce heavy snowfall over the Wasatch Mountains, but instead the moisture stays just to the north. In this instance, a small error in the position of the moisture results in a huge forecast bust.

For this reason, **ensemble modeling** is now the rage. Because the atmosphere is chaotic, it cannot be predicted with precision like the phases of the moon. Ensemble modeling involves producing many forecasts (each forecast is called an **ensemble member**) with different models or slightly different estimates of the initial structure of the atmosphere to predict a range of possibilities for the future. The National Centers for Environmental Prediction runs two ensemble modeling systems. One is the **Short-Range Ensemble Forecast System (SREF)**, which as of 2022 consisted of twenty-six members producing forecasts at sixteen-kilometer grid spacing for North America four times a day out to eighty-seven hours. The other is the **Global Ensemble Forecast System (GEFS)**, which as of 2022 consisted of thirty-one members producing global forecasts four times a day out to sixteen days (one run is extended to thirty-five days). The GEFS is often combined with the Canadian ensemble to create the **North American Ensemble Forecast System (NAEFS)**, which in aggregate has fifty-two members.

Ensemble modeling systems like the SREF and NAEFS can be used to generate event probabilities and examine a range of possible forecast outcomes. For instance, my group at the University of Utah uses NAEFS forecasts to produce forecasts for ski areas like Alta with snow accumulations from all fifty-two

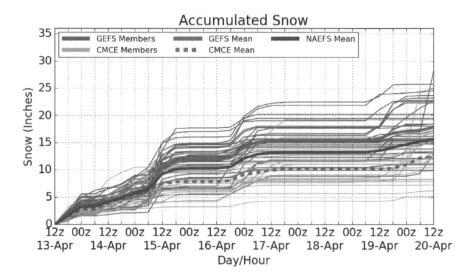

Figure 9.18. Plume diagram summarizing the total accumulated snowfall forecast for Alta from each member of the NAEFS (which combines the GEFS and Canadian ensembles [CMCE]) from 12 UTC (Z) 13 April (6 a.m. Mountain Daylight Time) to 12 UTC (Z) 20 April (6 a.m. Mountain Daylight Time) 2022. In this case, the driest member calls for five inches of snow and the wettest twenty-eight inches, illustrating the range of possible outcomes during the period. NAEFS forecasts have been adjusted to account for local terrain effects.

members displayed in a "plume" diagram (figure 9.18). Increasingly, **machine learning** is being used to improve the forecasts and probabilities obtained from ensemble modeling systems. It is likely that such efforts will improve forecasts dramatically in the coming years.

PUTTING IT ALL TOGETHER

By now, your head may be spinning. Where to get started with all these observations and computer model forecasts? I usually start with the past and present weather. Using the forecast funnel, I evaluate the movement of major weather systems and their impact on weather in the Wasatch Mountains. Then I produce a short-term forecast for the next several hours. I often make this forecast initially without consulting computer models before considering forecasts by the HRRR.

Beyond the short term, my forecast depends increasingly on computer model guidance. This usually involves examining the various computer model forecasts and adjusting them for common errors and local conditions. For example, during a period of unstable northwesterly flow, the models often fail to produce enough precipitation in the Cottonwood Canyons, so I'll modify the model forecasts to expect more snow in the Cottonwoods.

There is one other critical step in forecasting that is easily overlooked: the postmortem. A good forecaster evaluates not only the accuracy of his or her forecasts but also the factors that contribute to forecast success or failure. It is through continual self-evaluation that forecasters improve.

FINDING THE WEATHER DATA

Now that you are ready to become a forecaster, where can you find all this weather data? For mountain weather observations from the Wasatch Mountains, try wasatchsnowinfo.com/Wasatch and mesowest.utah.edu. For computer model forecasts, try weather.utah.edu. For pretty much everything else, tropicaltidbits.com/analysis/models/, twisterdata.com, or weather.rap.ucar.edu enable one-stop shopping for satellite, radar, surface observations, upper-air observations, and computer model forecasts, including help on how to interpret graphics. There are also subscription sites such as weathertap.com.

These and other websites provide a **firehose** of meteorological data and imagery. Good meteorologists learn to sip from this firehose by concentrating on those products and analyses that are most important. In *Blink: The Power of Thinking without Thinking*, Malcolm Gladwell calls this **thin slicing**. As part of your forecast postmortems, identify what products are the most useful and learn how to thin slice.

Forecasting the Greatest Snow on Earth

You now have all the tools needed to forecast the Greatest Snow on Earth. There's a learning curve, as in skiing or snowboarding; you won't be an expert on your first try. When I teach forecasting classes, we start forecasting right away, as doing is the best way to learn. After all, an expert is someone who has already made all the mistakes.

NOTE

1. A knot is about 15 percent faster than a mile per hour. For example, 10 knots equals 11.5 miles per hour.

10 Withering Winters

Let me let you in on a poorly kept secret. **Global warming** is real. The climate of the Wasatch Mountains today is warmer than it was when they were mining silver at Alta in the late nineteenth century. It is warmer than it was when lift-served skiing first came to Utah in the middle of the twentieth century. And it is warmer than it was in 1971 when Snowbird first opened the tram. We live in a different climate than the Mormon pioneers or even Ted Johnson and Dick Bass when they developed Snowbird in the late 1960s and early 1970s. How different? Let's have a look.

OUR WARMING WORLD

From the late 1800s to the 2010s, the Earth's average surface temperature increased about 1.8°F (figure 10.1). Imposed on this long-term trend are ups and downs of about 0.3°F that scientists call **climate variability**. Some of this variability is produced by the seesaw between El Niño and La Niña conditions in the tropical Pacific Ocean. El Niño warms the tropical Pacific Ocean, leading to a temporary increase in the Earth's average surface temperature, whereas La Niña cools the tropical Pacific Ocean, leading to a temporary decrease in the Earth's average surface temperature. Today, however, these ups and downs occur in a climate that is substantially warmer than that of the twentieth century. Globally, the eleven warmest years on record (as of 2021) have occurred since 2005, and the period from 2015 to 2021 was 0.5°F hotter than the previous seven-year period (2008 to 2014).

The temperature change since the middle of the twentieth century when skiing was beginning to take off as a winter sport varies regionally. It is greatest in the Arctic and over land compared to water (figure 10.2). Many mountain regions known for their skiing are warming faster than the global average including western North America, the European Alps, and Scandinavia. None of this is good for ice, snow, or skiing.

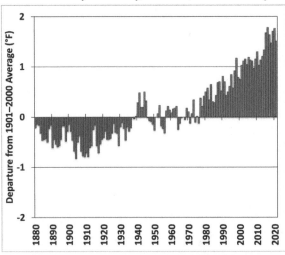

Figure 10.1. Departure (°F) of the Earth's annual average surface temperature from the twentieth century (1901–2000) average from 1880 to 2021. Source: National Center for Environmental Information.

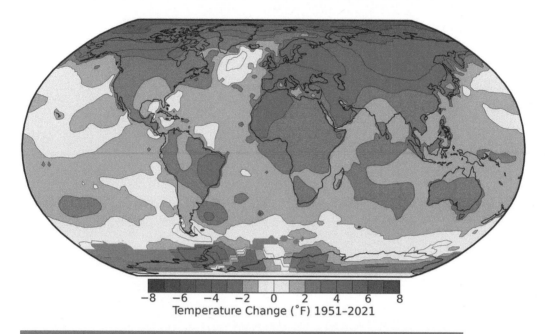

Temperature Change (°F) 1951–2021

Figure 10.2. Change in average surface temperature (°F) from 1951 to 2021. Gray areas indicate missing data. Source: NASA Goddard Institute for Space Studies Surface Temperature Analysis GISTEMP v4 (GISTEMP Team 2022, Lenssen et al. 2019).

DECLINING SEA ICE AND ICE SHEETS

The great melt has begun, and we are seeing major declines in the Earth's ice. The Arctic Ocean has historically been covered by **sea ice** for much of the year, with coverage greatest in March and smallest in September. However, satellite observations of Arctic **sea-ice extent**, a measure of the coverage of sea-ice over the Arctic Ocean and adjoining high-latitude water bodies, reveal a steady long-term decline since satellite observations began in 1979 (figure 10.3). The March sea-ice extent has declined at a rate of about 2.5 percent per decade, whereas in September, the decline is about 13 percent per decade. The thickness of Arctic sea ice is also decreasing. The melting of Arctic sea ice doesn't contribute to sea level rise, but it exposes a dark ocean that absorbs more sunlight, accelerating greenhouse-gas-induced warming.

Ice sheets are continental glaciers covering more than 50,000 square kilometers. The Greenland and Antarctic ice sheets are by far the largest, contain 99 percent of the world's fresh-water ice, and if fully melted would raise sea level by about 23 and 190 feet, respectively. In recent decades, the Greenland Ice Sheet has lost mass at an accelerating rate and is currently the biggest con-

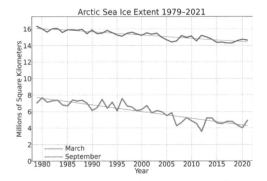

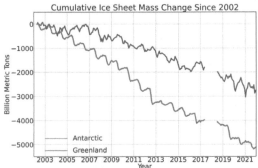

Figure 10.3. March (thick blue line) and September (thick red line) Arctic sea-ice extent from 1979 to 2021. Thin lines indicate linear trends. Source: National Snow and Ice Data Center (Fetterer et al. 2017).

Figure 10.4. Cumulative mass change of the Antarctic (blue) and Greenland (red) ice sheets since 2002. Satellite data gap from 2017 to 2018. Source: NASA Jet Propulsion Laboratory (Wiese et al. 2019).

tributor to sea-level rise. Observations from the NASA Gravity Recovery and Climate Experiment (GRACE) satellites have provided detailed observations of the Greenland Ice Sheet since 2002 and show a clear annual cycle with snowfall accumulation increasing mass during the cool season and melting and glacier calving (ice breaking off the glacier at its terminus, typically into the ocean or sea) decreasing mass during the warm season (figure 10.4). The warm season losses exceed the cool-season accumulation, resulting in a decline of about 280 billion metric tons of ice a year, the equivalent of 0.3 inches of sea level rise per decade. GRACE and other observations also indicate that the Antarctic Ice Sheet is losing mass, although at a slower rate than the Greenland Ice Sheet.

MELTING GLACIERS

Glaciers are iconic visual barometers of climate change that are melting at an alarming rate. From 1976 to 2016, glacial ice declined in all but one of the world's glaciated regions (figure 10.5). Losses were greatest in Alaska, where glaciers lost more than 3,000 billion metric tons of ice, which equates to about 780 cubic miles of ice volume, enough to raise sea level by about 0.4 inches. The only region with increasing glacier ice was the western portion of high-mountain Asia, including the Karakorum. The causes of this Karakorum anomaly are not fully understood, but it is not expected to persist with continued warming.

The South Cascade Glacier in the North Cascades of Washington has the longest continuous record of direct measurements in the United States, dating

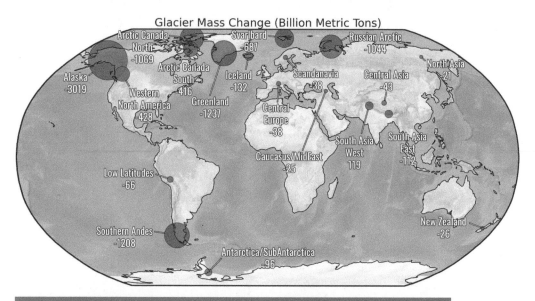

Glacier Mass Change (Billion Metric Tons)

Figure 10.5. Change in total regional glacier mass from 1961 to 2016. Red indicates mass loss and blue mass gain. Change in Greenland considers only glaciers distinct from the ice sheet. Adapted from Zemp et al. (2019). Source: World Glacier Monitoring Service.

back to the 1950s. It is one of forty-two reference glaciers recognized by the World Glacier Monitoring Service for long-term, high-quality observations. From 1955 to 2020, the South Cascade Glacier lost mass in sixty-four years and gained mass in only twenty years (figure 10.6). Glacier mass change is often expressed as an equivalent depth of water if the ice was melted and distributed over the area of the glacier. The cumulative mass loss of the South Cascade Glacier during this period was 38,360 millimeters, or about 125 feet. Photos illustrate the staggering loss of glacial ice including the dramatic retreat of the glacier toe and declining glacier thickness at upper elevations.

The loss of glacier ice has greatly affected resorts that offer glacier skiing during the summer. In North America, Whistler-Blackcomb used to operate two surface lifts on the Horstman Glacier for summer skiing, but one was removed in 2020 due to glacier-melt issues. Timberline in Oregon provides summer skiing on the Palmer Snowfield (technically not a glacier) and typically attempts to stay open until early September but was forced to close early in the summer of 2021 due to record-setting heat in the Pacific Northwest.

Summer skiing became popular in the European Alps in the 1970s, and by 1985 there were thirty-two operating glacier-skiing resorts, but dramatic gla-

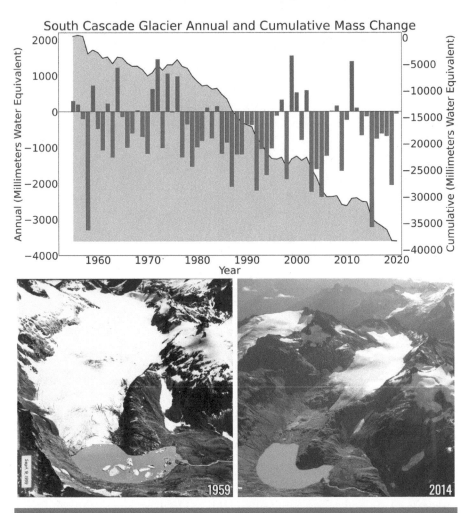

Figure 10.6. Annual and cumulative mass change of the South Cascade Glacier of Washington's North Cascades from 1955 to 2020 (top). Photos of the South Cascade Glacier in 1959 (bottom left) and 2014 (bottom right). Data source: World Glacier Monitoring Service (WGMS 2021). Photo source: USGS.

cier retreats and mass loss have occurred in recent decades. By 2021, only a handful of summer-skiing resorts remained. Hintertux (Austria) and Zermatt (Switzerland) still operate year-round with summer skiing on upper-elevation glaciers. Passo Stelvio (Italy) and Saas Fe (Switzerland) open for summer skiing but close for parts of the year. Finally, Kitzsteinhorn (Austria), Mölltal (Austria) Les 2 Alpes (France), Tignes (France), Val d'Isere (France), and Cervinia (Italy) operate for portions of July and/or August.

Figure 10.7. Glacier covers and snow farming operations during summer operations at Hintertux Glacier, Austria. Source: 80–20/Shuterstock.com.

In recent years, these resorts have employed snowmaking, wintertime water injection (creates a hard surface that reduces vulnerability to wind and other snow erosion influences), glacier covers (tarps that are white, reflect sunlight, and reduce snowmelt), and snow farming to preserve snow and enable summer operations (figure 10.7). Still glacier melt and retreat have in many cases reduced summer operations, the area of skiable terrain, and access to and from lifts that were designed for previous glacier coverage.

REGIONAL CLIMATE CHANGE AND SNOW

The contiguous western United States is warming faster than the global average, and the ten-year period ending in 2021 was 2.4°F warmer than the twentieth-century average temperature (figure 10.8). The region hasn't seen a year with a temperature below the twentieth-century average since 1993! Regional temperatures are dependent on the whims of the jet stream and weather systems, but thanks to global warming, the dice are now loaded for warm years and winters. A "cold" year in the early twenty-first century, such as

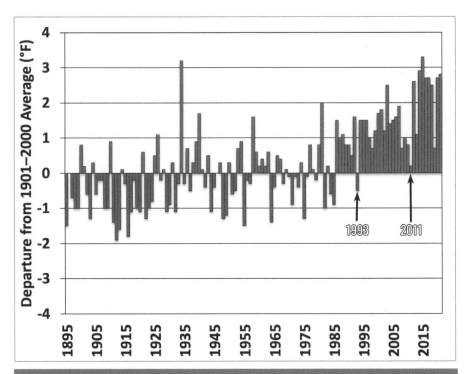

Figure 10.8. Departure (°F) of the annual average surface temperature in the contiguous western United States (Rockies westward) from the twentieth century (1901–2000) average from 1895 to 2021. Source: National Center for Environmental Information.

2011, would have been a near-average year when they first strung up lifts in the Wasatch Mountains during the middle of the twentieth century.

Identifying trends in many snow measures such as snowfall and snowpack is complicated by a lack of historical observations in mountainous regions. However, in the western United States, the Natural Resources Conservation Service has collected snow water equivalent observations of the snowpack at many mountain locations for decades. These observations are collected on

Figure 10.9 (facing page). Trend in April 1 snowpack snow water equivalent (percent) at observing sites in the contiguous western United States from 1955 to 2020 displayed geographically (top) and with elevation (bottom). Average trend at sites in each 2,000-foot elevation range (2,000–4,000, 4,000–6,000, 6,000–8,000, 8,000–10,000, and 10,000–12,000 feet) indicated in yellow boxes. Source: USDA Natural Resources Conservation Service and Environmental Protection Agency.

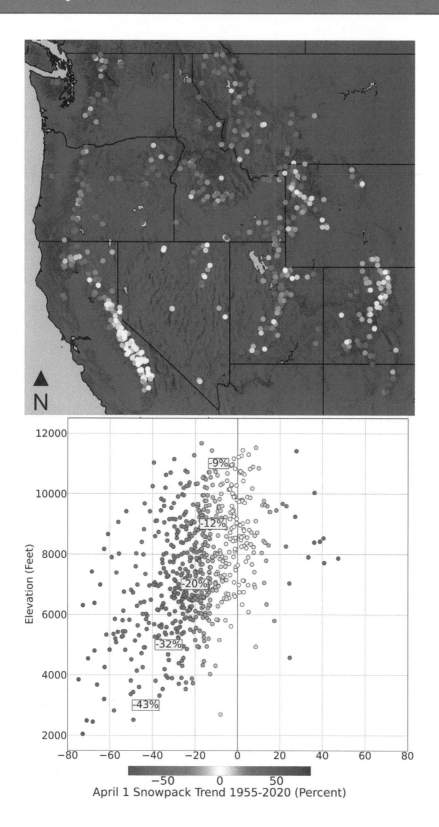

April 1 Snowpack Trend 1955-2020 (Percent)

or near April 1 when the snowpack is near peak depth. From 1955 to 2020, 86 percent of these sites exhibited a long-term decline in snow water equivalent (figure 10.9). The size of this decline is elevation dependent and on average is largest at sites between 2,000 and 4,000 feet, most of which are in the Pacific Northwest, and smallest at sites between 10,000 and 12,000 feet, most of which are in the southern Sierra Nevada and Colorado.

Many factors affect the April 1 snowpack at a given site including the temperature and precipitation during individual storms and the snow accumulation season, wind, aspect, vegetation and tree cover, cloud cover, and the presence of impurities like dust (which enhances the absorption of sunlight and snowmelt). Measuring snow is also challenging and has uncertainties. Thus, there is considerable variability in the trends from site to site. Nevertheless, research indicates that warming strongly contributes to the negative trends at low elevation sites in the Pacific states (Washington, Oregon, and California) where the snow climate is milder and more vulnerable to the initial increase in temperature associated with global warming. In contrast, the smaller negative trends at upper-elevation sites largely reflects declines in cool-season precipitation, which may not reflect the influence of global warming. The snow climate at these sites is colder and less vulnerable to recent temperature increases because wintertime temperatures are well below freezing.

Other ski regions are observing similar trends. For example, roughly two-thirds (68 percent) of snow stations in Austria observed negative trends in the average November to April snow depth from 1961 to 2021 (figure 10.10). No stations reported positive trends. For snow cover duration, 80 percent of stations observed shorter periods of snow cover, with declines largest at low-elevation stations where vulnerability to warming is the greatest.

CAUSES OF GLOBAL WARMING

Scientists first recognized the warming influence of **greenhouse gases** more than one hundred years ago. The natural **greenhouse effect** produced by the atmosphere raises the Earth's temperature by almost 60°F and creates the conditions for life as we know it. Since the beginning of the industrial revolution in the mid-eighteenth century, carbon dioxide concentrations have risen from 280 to 420 parts per million (ppm) due to fossil fuel combustion and deforestation and are now higher than at any time in at least the last two million years. More than half of this increase has occurred since the mid-1970s. Concentrations of other greenhouse gases, such as methane and nitrous oxide, are also increasing

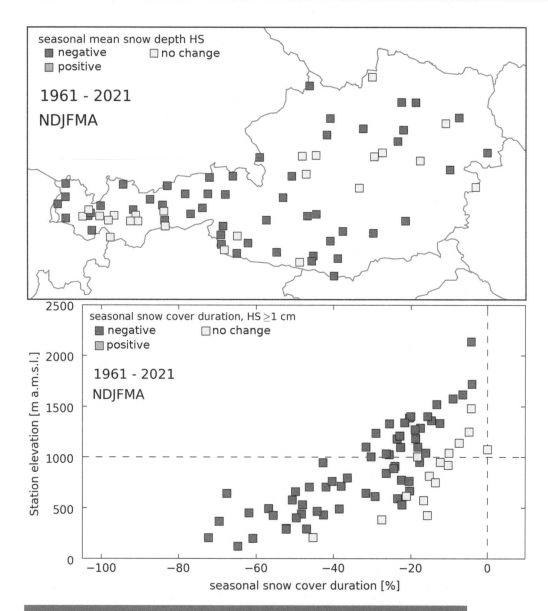

Figure 10.10. Trends in average November to April snow depth (top) and snow cover duration (bottom) at Austrian observing sites. Red and blue indicate statistically significant negative and positive trends, respectively (95 percent confidence level), although there are no sites in the latter category. Grey indicates no significant change. Station elevations in bottom panel in meters above mean sea level. Adapted from Olefs et al. (2021) with data updated through 2021.

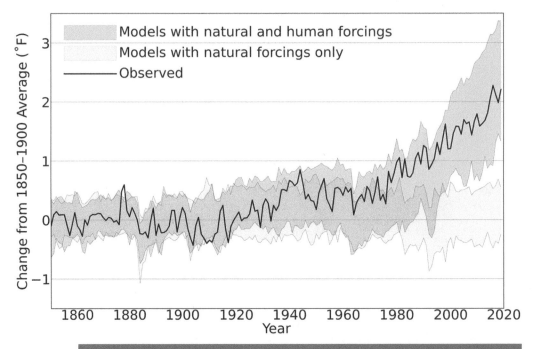

Figure 10.11. Comparison of the observed globally averaged surface temperature from 1850 to 2019 (black line) with a range of globally averaged surface temperatures produced by climate models run with only natural climate forcings (light orange) and with both natural and anthropogenic climate forcings (light blue). Data source: Gillett et al. (2021) and IPCC (2021).

due to human activities. These increases enhance the Earth's natural greenhouse effect and lead to **feedbacks** that further amplify the warming. For example, as temperatures rise, the water-vapor content of the atmosphere increases, which enhances the warming since water vapor is also a greenhouse gas.

More than 97 percent of actively publishing climate scientists conclude that most of the global warming in recent decades is caused by human activity, especially the buildup of greenhouse gases. Several lines of evidence support this conclusion. First, evidence of past climate provided by ice cores, tree rings, and other indicators illustrates that the warming in the past few decades is unprecedented in at least the last 2,000 years and possibly much longer. Second, meteorological observations show that the lower atmosphere has warmed, whereas the upper atmosphere has cooled, which is consistent with an enhanced greenhouse effect rather than a natural process such as an intensification of the output from the sun. Finally, the computer models that scientists use to simulate and understand climate change can reproduce the warming observed in the late twentieth and early twenty-first centuries only if they include anthropogenic effects such as the increase in greenhouse gases produced by human activities (figure 10.11). These climate models are unable to reproduce the warming if they include only known natural forcings. Bottom line: global warming is caused by humans.

11 The Fate of Skiing

In Greek mythology, three goddesses known as The Fates determine a person's life and destiny. Clotho spins the thread of life for mortal birth, Lachesis measures the thread to determine lifespan, and Atropos cuts the thread and determines the method of death.

In modern times, humans are The Fates of the climate system. We spun the thread and gave birth to global warming, and we will determine the destiny of the climate system depending on the path we take for future greenhouse gas emissions. Ultimately, that path will determine if skiing has a viable future or becomes increasingly confined to cold regions and high-altitude mountain areas.

TEMPERATURE PROJECTIONS FOR THE TWENTY-FIRST CENTURY

Projections of future climate change use multiple approaches including computer modeling. The computer models used to project future climate change simulate interactions between the atmosphere, oceans, snow and ice, and land surface (figure 11.1). Although similar to the computer models used to forecast the weather, climate models are not designed to predict the weather on a particular day but instead the average weather conditions over periods of many years. Confidence in the ability of climate models to provide credible estimates of future climate change is based on their ability to simulate the climate of the twentieth century. More than twenty research groups from around the world develop and maintain climate models, enabling a multimodel perspective of future climate change.

For twenty-first-century projections, different **scenarios** are developed using different levels of greenhouse gas emissions. In addition to projections for future emissions of carbon dioxide, methane, and other long-lived greenhouse gases, these scenarios also consider changes in air pollution (also called aerosols) and land-surface use and development.

Five scenarios were used for the latest assessment issued by the **Intergovernmental Panel on Climate Change (IPCC)**: (1) a very low greenhouse gas emissions scenario in which carbon dioxide emissions are cut to net zero around 2050 after which they become negative; (2) a low greenhouse gas emissions scenario in which carbon dioxide emissions are cut to net zero around 2075 after which they become negative; (3) an intermediate greenhouse gas emissions scenario in which carbon dioxide emissions stabilize near current levels in 2050 and then decrease but do not reach net zero by 2100; (4) a high greenhouse gas emission scenario in which greenhouse gas emissions double by 2100; and (5) a very high greenhouse gas emissions scenario in which carbon dioxide emissions triple

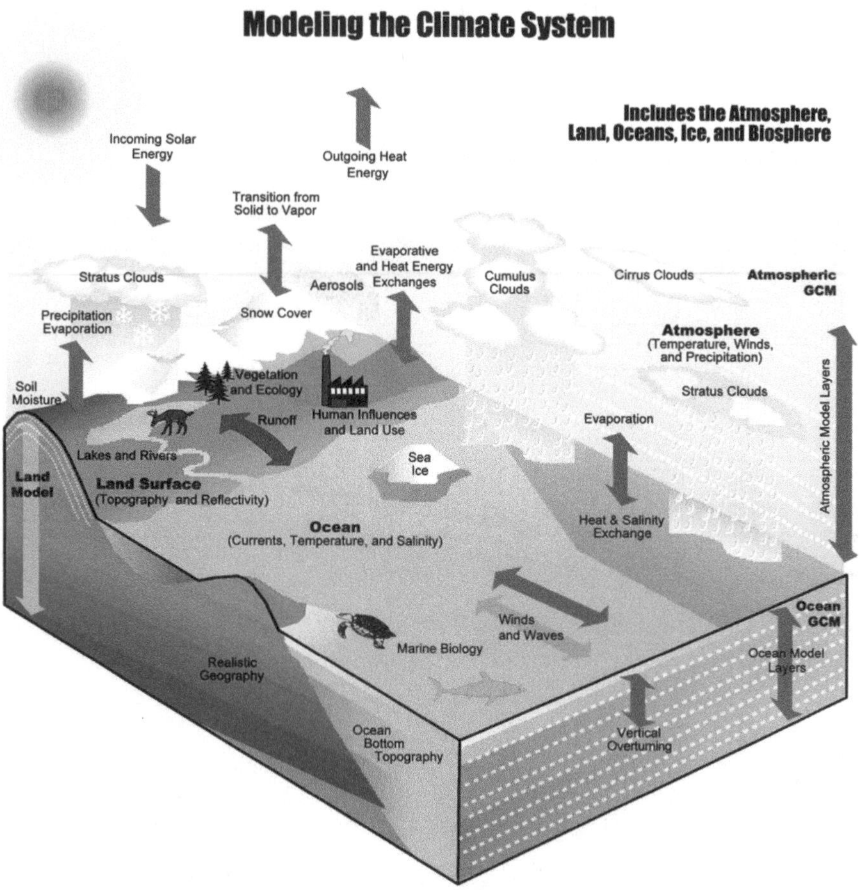

Figure 11.1. Major components of the climate system that are considered in climate simulations. From Karl and Trenberth, SCIENCE 302: 1719 (2003).

by 2075. We will refer to these as very low, low, intermediate, high, and very high, respectively. Negative emissions in the very low or low scenario in the late twenty-first century require deep emissions cuts, enhancement of natural processes that remove greenhouse gases from the atmosphere, and/or the development of technologies to capture and store greenhouse gases from the atmosphere.

For all five scenarios, global temperatures are expected to increase through mid-century (2041 to 2060; figure 11.2). For the near future (2021 to 2040), there is actually little difference between the five scenarios since we are already committed to some additional warming as the Earth continues to adjust to prior greenhouse gas emissions. Relative to the average temperature from 1850 to 1900, the best estimates for near future (2021 to 2040) warming under the five

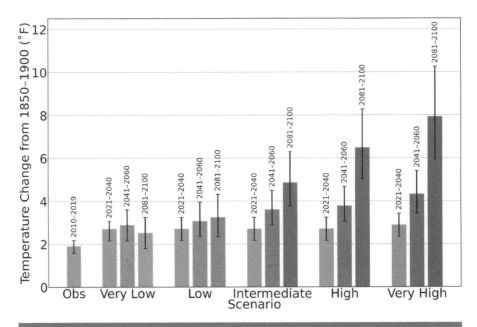

Figure 11.2. Observed (Obs, for 2010–2019) and projected global surface temperature change from 1851–1900. Projections based on very low, low, intermediate, high, and very high scenarios for the near-term (2021–2040), mid-century (2041–2060), and late-century (2081–2100) periods. Bars represent best estimates. Capped lines denote the very likely range. Data sources: National Centers for Environmental Information, IPCC (2021).

scenarios are 2.5 to 2.7°F, which is another 0.6 to 0.8°F warmer than observed from 2010 to 2019. Estimates vary depending on the model. Capped lines in figure 11.2 illustrate the very likely range for each scenario and period.

Warming after 2040 strongly depends on the future emissions path we take. In the very low emissions scenario, best-estimate temperatures peak mid-century at about 2.7°F warmer than 1850 to 1900 before declining slightly in the late-century (2081 to 2100). In contrast, in the very high emissions scenario, best-estimate temperatures climb through the century and by late-century are 7.9°F warmer than 1850 to 1900. Sadly for skiers, the very low emissions scenario is increasingly unlikely as even current pledges for greenhouse gas reductions are insufficient to take that path. Even the low emissions path is looking unlikely. Some have argued the most likely scenario is the intermediate one. Under this scenario, best estimate temperatures increase to 3.6°F warmer than 1850 to 1900 by mid-century and 4.9°F warmer by late century. These increases are 1.7°F and 3.0°F warmer than observed from 2010 to 2019.

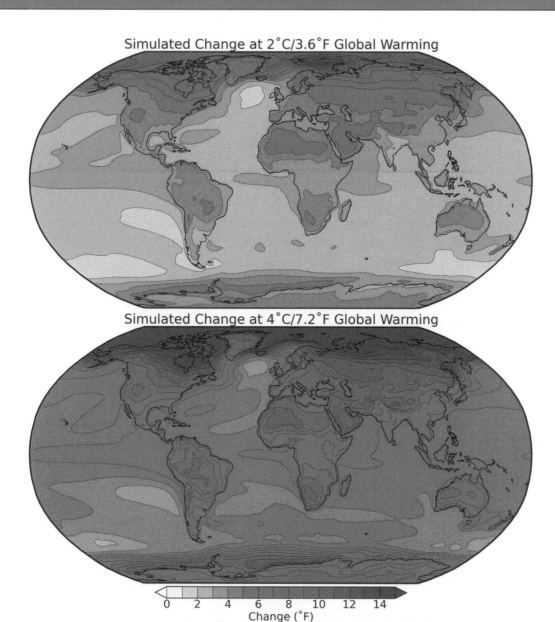

Simulated Change at 2°C/3.6°F Global Warming

Simulated Change at 4°C/7.2°F Global Warming

Figure 11.3. Simulated change in annual mean temperature for a global surface temperature increase of 2°C/3.6°F (top) and 4°C/7.2°F (bottom) relative to 1851–1900. Data source: Fischer and Hauser (2021) and IPCC (2021).

The spatial pattern of projected warming is similar to that observed during the twentieth century and remains consistent in structure as the average global temperature increases (figure 11.3). The greatest warming occurs in the Arctic, over land compared to ocean, and over the interior of continents compared to the coasts. For a 4°C/7.2°F increase in average global temperature relative to 1850 to 1900, the interior western United States and northern Utah warm more than 10°F, comparable to the current difference in average annual temperature between Salt Lake City and Park City.

IMPACTS OF WARMING ON THE GREATEST SNOW ON EARTH

Based on the climate model projections discussed above and other research on global and regional climate, there is high confidence that Utah will warm during the twenty-first century, with the rate and size of the temperature rise depending largely on future greenhouse gas emissions. The whims of the jet stream and weather systems will continue to cause year-to-year and decade-to-decade variations in temperatures, precipitation, snowfall, and snowpack, but global warming is already altering our snow climate and will exert a growing influence on the Greatest Snow on Earth through at least mid-century. After that, much will depend on what emissions path we follow.

More Rain, Less Snow

Global warming is already causing a greater fraction of wintertime precipitation to fall as rain instead of snow at lower elevations in northern Utah, and this trend will accelerate and extend into the higher elevations with continued warming. Estimates of how much of the wintertime precipitation that fell as snow during the period from 1979 to 2008 would instead fall as rain if winter storms were 1°C, 2°C, 3°C, or 4°C warmer illustrate this well (figure 11.4). For a temperature rise of 1°C (about 1.8°F), about 10 percent more wintertime precipitation would fall as rain at 7,000 feet (roughly the base elevation of Park City Mountain Resort and Deer Valley). At 9,500 feet (mid-mountain at Snowbird and Alta and upper mountain at Park City Mountain Resort and Deer Valley), however, it's only 3 percent. This reflects the fact that the upper elevations are well above the freezing level during most storms and thus have some "insurance" against global warming.

The numbers worsen with greater warming. For a 4°C temperature increase (about 7.2°F), about 40 percent more wintertime precipitation would fall as rain at 7,000 feet. At 9,500 feet, it's about 20 percent. These elevation-dependent effects mean that the lower elevations will suffer the most and upper-elevation ski terrain will become an increasingly precious and valued commodity for skiing.

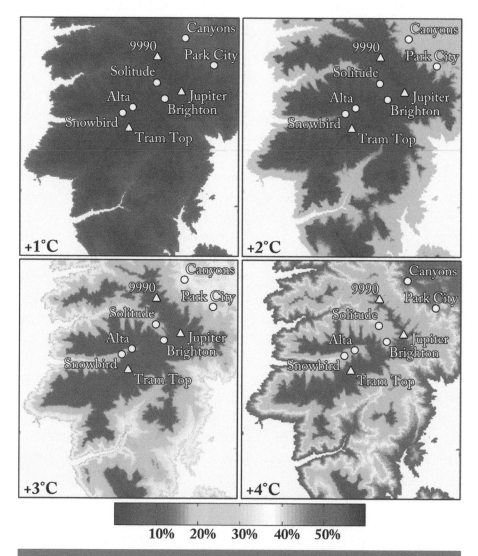

Figure 11.4. Percent of wintertime precipitation that currently falls as snow that would instead fall as rain for a temperature increase of 1°C, 2°C, 3°C, and 4°C. Selected ski-area bases (circles) and summits (triangles) annotated. Adapted from Jones (2010).

A second likely consequence of global warming is a decrease in snow quality. The water content of freshly fallen snow at Alta is closely related to temperature. Colder storms typically produce lower water content snow, but these storms will become less common in a warming climate. Based on current relationships between temperature and water content, 4°C (7.2°F) of warming would increase the average water content of snow at Alta from 8.4 percent to 10.0 percent.

Changes to Storms and Storm Tracks

The future snow climate of the Wasatch Mountains will also be influenced by changes to storm characteristics and storm tracks. Will these changes increase or decrease wintertime precipitation in the Wasatch Mountains? Will they help offset or amplify losses in snowfall and snowpack because of global warming? These questions are difficult to answer.

Averaged across all the climate models, the average annual precipitation increases in the high latitudes and decreases in many dry subtropical regions (figure 11.5). This pattern remains consistent as the average global temperature increases, with wet areas getting wetter and dry areas getting drier. At a 4°C / 7.2°F increase in average global temperature relative to 1850 to 1900, the average annual precipitation in the interior western United States, including Utah, is about 15 percent greater. However, there is not a lot of confidence in this change because there isn't strong agreement amongst the climate models, which produce a wide range of trends for Utah's precipitation, including some that call for a decrease.

If precipitation were to increase, it might help offset some of the snowfall losses due to a greater fraction of precipitation falling as rain instead of snow. However, if precipitation were to decrease, it would further exacerbate declining snowfall and snowpack trends due to global warming. Another challenge in this area is the inability of current climate models to resolve the effects of the Wasatch Mountains and Great Salt Lake. Ideally, we would run higher-resolution climate models, but this is not practical with current computer systems. This, combined with the lack of consensus among the climate models, makes future trends in the annual and wintertime precipitation over northern Utah and the Wasatch Mountains difficult to assess with confidence.

Snowpack Trends

Snowpack trends result from the combined effect of temperature trends, which affect the fraction of precipitation that falls as snow and snowmelt, and precipitation trends. As you might surmise from chapter 10 and the discussion above, the outlook for the Wasatch Mountain snowpack is not rosy. During the twenty-first century the insidious effects of a warming climate will become increasingly clear. More precipitation will fall as rain instead of snow, the length of the snow accumulation season will decrease, and there will be more snowmelt events during the winter. These effects will be most pronounced in

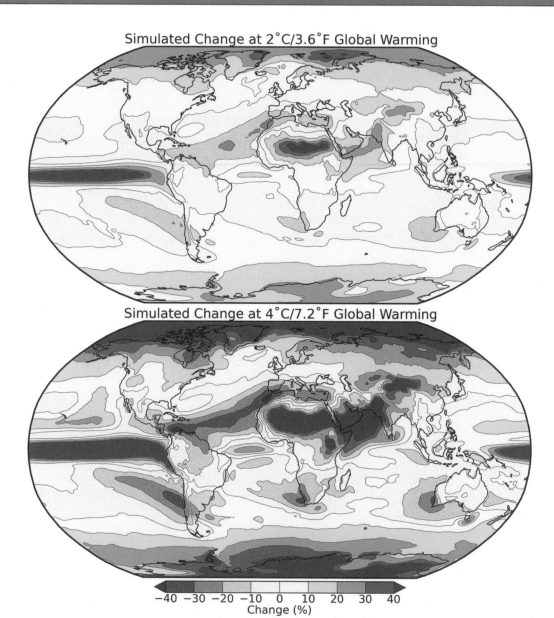

Figure 11.5. Simulated change in annual mean precipitation for a global surface temperature increase of 2°C/3.6°F (top) and 4°C/7.2°F (bottom) relative to 1851–1900. Large positive percentage changes can correspond to small absolute changes in arid regions such as the Sahara Desert. Data source: Fischer and Hauser (2021) and IPCC (2021).

lower-elevation ski terrain. Declines will be smaller at upper elevations (above 9,000 feet) due to the colder climate, lower sensitivity to temperature change, and the possibility of increased wintertime precipitation.

THE FATE OF SKIING IN THE INTERIOR WESTERN UNITED STATES

Scientists have used computer modeling to estimate how global warming will affect the snow climate at ski resorts in the interior western United States and other regions of the world. These studies highlight the uneven influence of global warming on snowfall and snowpack, with resorts in warmer climates and at lower elevations generally experiencing the greater losses than resorts in colder climates and at higher elevations.

For example, scientists have used computer modeling to compare the historical (1981 to 2011) snow climate at ski resorts in the interior western United States to one with similar storm-track characteristics but regional temperatures that are about 2.0°C/3.6°F warmer. Under this scenario, the median number of "natural" core-season ski days, defined as the number of days between 15 November and 15 April with a natural snowpack containing at least eight inches of snow water equivalent at an elevation just below mid-mountain, declines at every ski resort (figure 11.6). The largest declines are at lower elevation resorts in Idaho and Montana and more southern resorts in New Mexico. Averaged across all ski resorts, the median natural core-season length decreases from 107 to 76 days and only twenty-two resorts have more than one hundred natural ski days, which is sometimes used as an indicator of business viability. These changes reflect a later start to the snow accumulation season in the fall, a decline in peak snowpack, and an early and more rapid spring melt. Peak snowpack increases only occur at the highest elevations (above 10,000 feet) at a small number of ski resorts in Montana, Wyoming, and Colorado. In the central Wasatch, the natural core-season days decrease by twenty-five to fifty days, with the smallest declines at Alta and the largest at Sundance.

We can also examine other variables that affect resort operations and snow quality. The average number of "warm" days during the core season with an average temperature above 32°F at an elevation just below mid mountain increases at all resorts (figure 11.7). The largest increases are in Idaho and New Mexico, whereas the smallest increases are at high-altitude resorts in central Colorado. Days with at least 0.04 inches of rain at an elevation just below mid-mountain increases substantially at many resorts in Idaho and Montana and even at some resorts in northern Utah (figure 11.8). Higher altitude resorts in

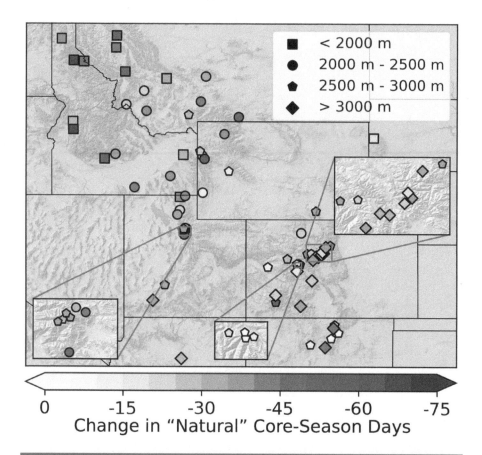

Change in "Natural" Core-Season Days

Figure 11.6. Change in the median number of "natural" core-season days (i.e., days between 15 November and 15 April with a natural snowpack that contains at least eight inches of snow water equivalent at an elevation just below mid-mountain) at interior west ski resorts compared to 1981 to 2011 if regional temperatures are 2.0°C/3.6°F warmer. Adapted from and based on computer modeling by Lackner et al. (2021). © American Meteorological Society. Used with permission.

Utah and especially Colorado see much smaller increases (some high-altitude resorts in Colorado see little to no change as these resorts remain cold enough that significant rain is extremely rare during the core season). Finally, there are decreases in the total early-season (October 1 to December 31) snowmaking production potential at all ski resorts (figure 11.9). These results indicate that ski-season struggles will be greatest for resorts in warmer, lower altitude or lower latitude regions of the interior western United States, whereas snow and skiing will be most resilient in colder, high-altitude areas, especially in Colorado. In Utah, resorts in the Cottonwoods are the least vulnerable to climate change.

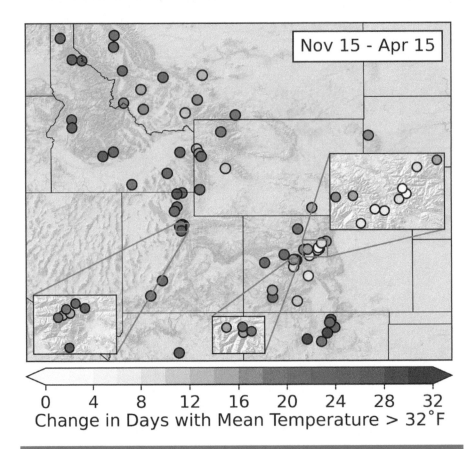

Figure 11.7. Change in the average number of "warm" core-season days (i.e., days between 15 November and 15 April with an average temperature above 32°F at an elevation just below mid-mountain) at interior west ski resorts compared to 1981 to 2011 if regional temperatures are 2.0°C/3.6°F warmer. Adapted from and based on computer modeling by Lackner et al. (2021). © American Meteorological Society. Used with permission.

Snowmaking will be of growing importance everywhere, although early season snowmaking production potential will decline.

A DIRTY LITTLE SECRET

Greenhouse gases get most of the attention in media coverage of climate change. There is good reason for this, as they are the 800-pound gorilla in the zoo and the largest driver of recent and future climate change. Nevertheless, there are other factors that affect the snow climate of the Wasatch Mountains and other mountainous regions, most notably dust.

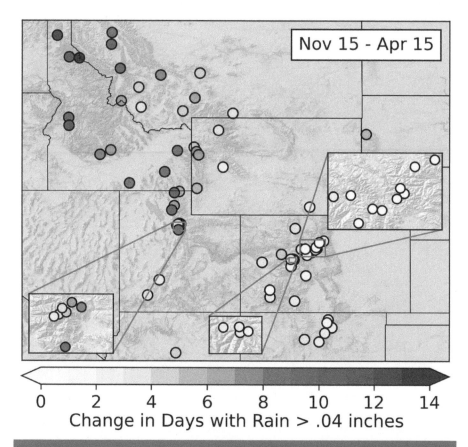

Change in Days with Rain > .04 inches

Figure 11.8. Change in the average number of "wet" core-season days (i.e., days between 15 November and 15 April with at least 0.04 inches of rain at an elevation just below mid mountain) at interior west ski resorts compared to 1981 to 2011 if regional temperatures are 2.0°C/3.6°F warmer. Adapted from and based on computer modeling by Lackner et al. (2021). © American Meteorological Society. Used with permission.

Dust storms frequent the Wasatch Mountains during the winter and spring, leaving dirty layers in the snow (figure 11.10). As the snow melts during the spring, water percolates through the snowpack and the dust collects on the surface. This leads to a remarkably dirty snowpack some call **snirt**—part snow, part dirt.

Dust has important implications for snowmelt. In Utah and many other mountainous regions, most of the energy that melts snow comes from the sun. Freshly fallen snow is very white and reflects about 90 percent of the sun's energy back to space. An older, relatively dust-free snowpack reflects about 70 percent of the sun's energy back to space. Dust, however, is much darker than

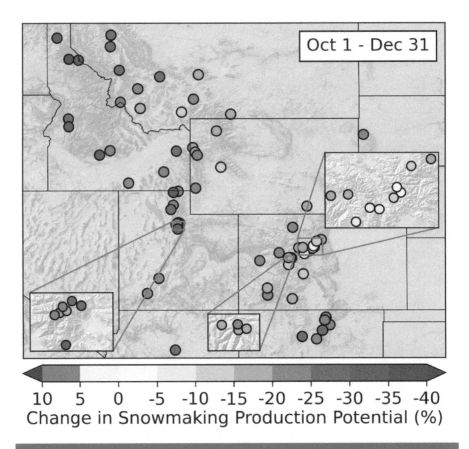

Figure 11.9. Change in the early season snowmaking production potential at an elevation just below mid-mountain at interior west ski resorts compared 1981 to 2011 if regional temperatures are 2.0°C/3.6°F warmer. Adapted from and based on computer modeling by Lackner et al. (2021). © American Meteorological Society. Used with permission.

snow. A snowpack that is laden with dust reflects only 35–50 percent of the sun's energy back to space (figure 11.11).

The energy that is not reflected back to space warms and melts the snowpack. At noon on a clear day in mid-May, about 100 watts of sunlight shines on the snowpack per square foot. Without dust, the snow absorbs only about 30 watts of this energy. With dust, however, it absorbs an additional 30 watts, doubling the energy available for snowmelt.

Therefore, the presence of dust accelerates the snowmelt. For a spring ski day, this can be advantageous when the snow is rock hard and you want it to

Figure 11.10. Layers of dust in a spring snowpack near Alta. Courtesy Annie Burgess.

soften up, but it can also lead to a shorter period of good corn snow, which turns quickly into mashed potatoes. Later in the spring, dust leads to an earlier loss of snow cover. Studies in the San Juan Mountains of Colorado suggest that dust decreases the duration of snow cover by thirty to fifty days.

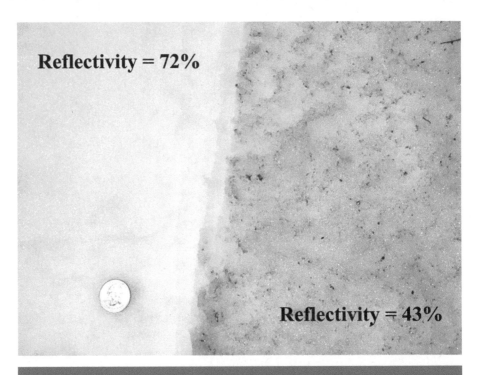

Figure 11.11. Comparison of the reflectivity of clean and dust-laden snow. Courtesy Thomas H. Painter, JPL/Caltech.

Most of the dust in the Wasatch Mountains comes from dry lake, agricultural, and desert regions of southwest Utah during periods of strong southerly or southwesterly flow (figure 11.12). The old lakebed of the shrinking Great Salt Lake is also a growing dust source for the region. Natural land surfaces in these regions include desert pavement, physical crusts, and **cryptobiotic soil** (figure 11.13), a biological soil crust found in deserts around the world. These surfaces are very resistant to wind erosion and dust emissions, but are easily disturbed by livestock grazing, vehicle traffic, and other human activities. Dust in the Wasatch Mountains is a consequence of emissions and transport from areas where the land surface has been disturbed.

I often call dust the dirty little secret of the Greatest Snow on Earth. In most years, you probably won't see it during the winter unless you get out a shovel and do some digging. In the spring, however, you can't miss it as the snow melts and the dust collects on the surface. Reducing the number and intensity of dust sources could be a way to mitigate the effects of global warming on snow, slow the spring snowmelt, and enable a longer ski season in a warming world.

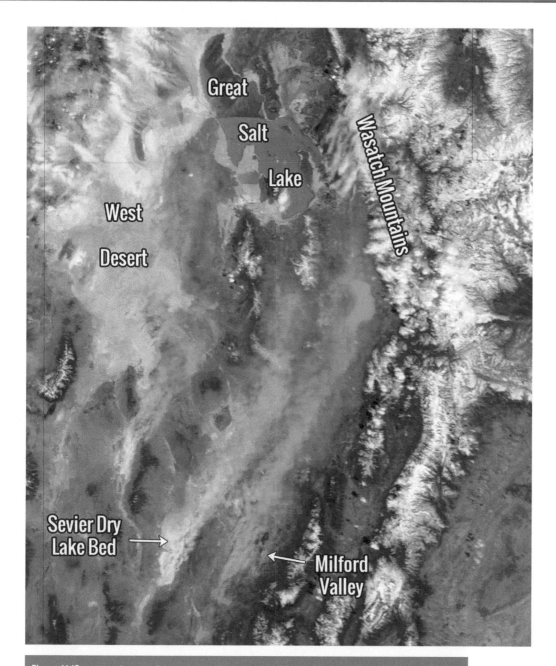

Figure 11.12. Dust plumes from the Sevier Dry Lakebed, Milford Valley, and other areas of southwestern and western Utah on April 19, 2008. Satellite image courtesy of NASA.

Figure 11.13. Cryptobiotic soil crust in Canyonlands National Park. Hiking trails illustrate the sensitivity of the crust to disturbance.

WILL WE SAVE SKIING?

Today we stand at a crossroads, and the path we choose will determine the fate of skiing. Some loss of snow in many ski regions is now unavoidable, but efforts to aggressively reduce greenhouse gas and dust emissions will enable skiing to still have a viable future, albeit one in which we may see shorter seasons, less reliable natural snowpacks, and the loss of resorts in vulnerable areas. Continued emissions, however, will lead to a future in which snow and skiing become increasingly confined to a few high-altitude mountain areas. We must act now to save skiing.

Glossary

aggregates. Interlocked snowflakes.

aggregation. The merger of two or more snowflakes.

Alpine rim. The outer edge of the European Alps.

anchors. Trees, rocks, and other features that help hold the snowpack in place.

aspect. The direction a slope faces.

atmospheric river. A narrow filament of atmospheric moisture that often originates in the tropics or subtropics.

avalanche paths. Areas along which avalanches travel.

avalanche danger rose. A graphical illustration of the avalanche danger by elevation and aspect commonly used in avalanche forecasts.

backcountry. Areas outside of ski-area boundaries where the snowpack is not subject to regular avalanche mitigation.

bed surface. The surface on which an avalanche slides.

Bergeron process. A mechanism for ice-crystal growth in mixed-phased clouds involving the net transfer of water molecules from cloud droplets to ice particles.

blocking front. The boundary between the approaching flow and air that is blocked or stagnated on the windward side of a mountain range.

blower pow. Very low water content snow.

bottomless. Powder in which the skis or snowboards float without riding on the underlying surface.

bright band. An area of high radar reflectivity associated with melting snow.

Cascade concrete. Wet, high water content snow that falls frequently in the Cascade Mountains.

chaff. Military countermeasures designed to generate large radar returns.

champagne powder. Low water content snow.

climate variability. Fluctuations from the long-term average climate.

climbing skins. Removable pieces of fabric that allow skis to glide forward but grip going backward.

cloud condensation nuclei (CCN). Vanishingly small particles on which condensation occurs to form cloud droplets.

cloud droplets. Tiny water droplets that comprise clouds.

cloud seeding. Adding artificial particles to the atmosphere or clouds to alter the amount or type of precipitation produced, including efforts to enhance precipitation or suppress fog and hail.

cloud storms. Clouds that produce virga but little precipitation at the ground.

cold smoke. Very low water content snow.

columns. Ice crystals with large prism facets and small basal facets.

computer models. Mathematical equations solved using computers to predict the weather and climate.

condensation. Water molecules transitioning from vapor to liquid.

continental snow climate. A snow climate featuring a shallow snowpack, lower temperatures, and less frequent storms that usually feature low water content snow and rarely produce rain.

convergence. The meeting of atmospheric flows.

country club day. A day when Alta and Snowbird are open, but the Little Cottonwood Canyon highway is closed.

crown. The fracture face at the top of a slab avalanche.

cryptobiotic soil. A soil crusted with cyano bacteria, algae, fungi, mosses, and other organisms.

dendritic growth region. An area in a cloud where the temperature and relative humidity favor the growth of dendrites.

deuterium. A "heavy" hydrogen atom that includes a neutron.

dewpoint. The temperature at which net condensation occurs if the air is cooled at constant pressure.

dreaded lake effect (DLE). Meteorological nickname for the Great Salt Lake effect.

dry avalanches. Avalanches involving predominantly frozen snow, typically at temperatures below freezing.

dust-on-crust. A shallow, low-density snowfall through which skis sink and ride on the underlying hardpack.

ensemble member. One forecast from an ensemble modeling system.

ensemble modeling. The generation of a suite of computer model forecasts.

facets. The faces (sides) of a prism.

FACETS. An acronym developed by Ian McCammon summarizing the human factors contributing to avalanche accidents (Familiarity, Acceptance, Commitment/Consistency, Expert halo, Tracks, Scarcity).

false alarms. Model forecasts that don't verify.

feedbacks. Linkages between processes than enable an initial forcing to amplify.

firehose. The torrent of weather data flowing to meteorologists.

föhn. A warm and potentially strong downslope wind in the European Alps and other mountainous regions.

forecast funnel. A forecasting technique that begins with larger weather scales and then funnels to smaller scales.

Gazex avalanche-mitigation systems. An elbowed steel tube that projects an oxygen/propane-based explosion onto the snow surface to trigger avalanches.

geostationary orbit. A satellite orbit in which the time required to perform one orbit matches that of one rotation of the Earth, enabling the satellite to remain above a fixed point on the Earth's surface.

glaciation. The process of a liquid cloud becoming a mixed-phase cloud consisting of both liquid and ice particles.

glaciers. Slowly moving perennial masses of ice formed over land by the accumulation and compaction of snow.

Global Ensemble Forecast System (GEFS). A modeling system run by the National Centers for Environmental Prediction that produces an ensemble of global forecasts.

Global Forecast System (GFS). A forecast model run by the National Centers for Environmental Prediction that produces global forecasts.

global warming. The rise in the Earth's temperature since the late nineteenth century.

Goldilocks storms. Storms that produce enough snow for bottomless powder without creating overly challenging trail-breaking or avalanche conditions.

Gosetsu Chitai. Japanese heavy snow region near the Sea of Japan.

graupel. A ball, cone, lump, or hexagonal-shaped snow pellet produced by the riming of a snowflake or ice particle.

Great Salt Lake effect. Precipitation generated by the influence of the Great Salt Lake.

greenhouse effect. The process by which atmospheric gasses absorb and emit infrared radiation, elevating the Earth's temperature compared to what it would be without the existence of greenhouse gasses.

greenhouse gases. Atmospheric gases that absorb and emit infrared radiation.

grid cells. Boxes and other shapes that define the stencil used to make mathematical calculations in weather and climate models.

grid spacing. The length of the side of a grid cell.

ground clutter. High radar reflectivities produced by mountains and other ground-based objects.

habit. The shape of an ice crystal.

hero snow. A right-side-up snowfall enabling effortless powder skiing.

hexagonal prism. The three-dimensional shape taken by ice crystals that includes six rectangular prism facets (or sides) with two hexagonal basal facets at the top and bottom.

High-Resolution Rapid Refresh (HRRR). A forecast model run by the National Centers for Environmental Prediction that produces forecasts at three kilometer grid spacing for the continental United States and Alaska.

ice multiplication. The rapid increase in the number of ice particles in a cloud as ice crystals shatter due to freezing and collisions.

ice nucleus. A small particle that helps water arrange its molecules into an ice crystal.

ice sheet. A mass of continental glacial ice covering an area of at least 50,000 square kilometers.

infrared (IR) satellite images. Satellite images based on infrared radiation in a band in which atmospheric gases are largely transparent (sometimes referred to as "window" or "clear" IR).

Intergovernmental Panel on Climate Change (IPCC). A body of the United Nations that assesses climate change–related science.

interlodge. A period of restricted travel during high avalanche hazard in upper Little Cottonwood Canyon.

land-breeze convergence. The meeting of land breezes from opposing shorelines.

land breezes. Winds that blow from land to water when the air over the water is warmer than the air over the land.

leeward. Facing away from the prevailing wind.

loose-snow avalanches. Avalanches in which loose snow slides down a slope, often creating a fanlike pattern.

machine learning. In meteorology, the use of computer algorithms to improve the use of weather data and computer model forecasts.

maritime snow climate. A snow climate featuring a deep snowpack, mild temperatures, and frequent storms that produce high water content snow and occasional rain.

maximum security interlodge. An interlodge period during which people must move into the areas of houses or lodges that are most protected from avalanches.

megadrought. A persistent drought event lasting for at least a decade.

meteograms. Graphs showing time series of weather observations.

microclimate. A unique local climate compared to the surrounding region.

midlake bands. Lake-effect precipitation bands that form near or along the long axis of the Great Salt Lake or other elongated bodies of water.

mixed-phase cloud. A cloud consisting of a mixture of supercooled water droplets and ice particles.

mixing cloud. A cloud that forms from the mixing of two air masses with differing temperatures and relative humidities.

North American Ensemble Forecast System (NAEFS). An ensemble modeling system that combines the Global Ensemble Forecast System (GEFS) from the United States and the Canadian ensemble.

North American Mesoscale Forecast System (NAM). A forecast model run by the National Centers for Environmental Prediction that produces forecasts for North America.

North American Public Avalanche Danger Scale. A scale used by Canadian and American avalanche forecasters to summarize the likelihood, size, and distribution of avalanches.

numerical weather prediction (NWP). The use of mathematical models to forecast the weather.

orographic convection. Cells or bands of strong rising motion and precipitation formed during flow over hills or mountains.

orographic precipitation. Precipitation generated or enhanced by hills or mountains.

plates. Ice crystals with short prism facets and large basal facets.

powder fever. The clouding of objective decision making by the desire for deep powder.

precipitation shadow. An area of reduced or no precipitation on the leeward side of a mountain barrier.

radar. A device used to detect precipitation and other objects using radio waves.

radar reflectivity. The intensity of radar power returned to the radar receiver.

refraction. The bending of a radar beam.

right-side-up snowfall. A snowfall with low water content snow on top of high water content snow.

rime. A coating of ice produced when supercooled cloud or drizzle drops freeze on objects like trees or chairlifts.

salinity. The dissolved salt content of water.

Satoyuki. A snowstorm generated by the Sea of Japan with accumulations greatest in the coastal lowlands.

saturation. When the atmosphere has a relative humidity of 100 percent.

scenarios. Projected future trends of greenhouse gas emissions, air pollutant emissions, and land-cover characteristics.

sea-effect snowstorms. Snowstorms produced during cold-air outbreaks over warm seas, such as the Sea of Japan.

sea ice. Ice that forms from saltwater freezing and that floats on the water surface.

sea-ice extent. A measure of the regional coverage of sea ice.

seeder-feeder. The growth of precipitation as it falls from seeder clouds aloft into low-level feeder clouds.

sidecountry. Backcountry that can be accessed from a lift with no climbing.

Sierra cement. Wet, high water content snow that frequently falls in the Sierra Nevada.

slab avalanche. An avalanche that begins as a cohesive slab or plate of snow.

slackcountry. Backcountry that can be accessed from a lift with some climbing.

sleet. Precipitation consisting of frozen raindrops.

sluffs. Small loose-snow avalanches.

snirt. Snow that is covered with dust.

SNOTEL (SNOpack TELemetry) stations. Weather and snowpack monitoring stations operated by the Natural Resources Conservation Service.

snow level. The altitude at which falling snow changes to rain.

snow-safety professionals. Individuals involved in avalanche study, education, prediction, and mitigation.

snow-study plot. An area where snow profiles are examined to assess snowpack structure and stability.

snow water equivalent. The amount of water in the snow expressed as a theoretical depth if the snow melted instantaneously.

spillover. Precipitation that is carried downstream and falls out on the leeward side of a mountain barrier.

Short-Range Ensemble Forecast System (SREF). A modeling system run by the National Centers for Environmental Prediction that produces an ensemble of forecasts for North America.

stability. The resistance of the atmosphere to vertical motions.

stellar dendrites. Snowflakes with six treelike arms.

sublimation. Water molecules transitioning from ice to vapor.

supercooled. Water that is below 32°F but remains unfrozen.

supersaturated. When the atmosphere has a relative humidity greater than 100 percent.

terminal lake. A lake with no outlet.

terrain concavity. A hollow in the topography.

terrain-driven convergence. The meeting of two or more atmosphere flows due to the influence of topography.

terrain traps. Areas where avalanche snow can pile up and deeply bury a skier.

thin slicing. Quick decision making while evaluating an overwhelming amount of data or information.

transitional snow climate. A snow climate intermediate to maritime and continental snow climates.

ultrasonic snow-depth sensor. An instrument that uses ultrasonic sound waves to measure snow depth.

upside-down snowfall. A snowfall with high water content snow on top of low water content snow.

vapor deposition. Water molecules transitioning from vapor to ice.

vertical profile of snow water content. The change in water content from the bottom to the top of the snow.

virga. Precipitation that evaporates or sublimates before reaching the ground.

visible satellite images. Satellite images based on visible radiation.

water content. The percentage of the snow that is frozen or liquid water.

water vapor images. Satellite images based on infrared radiation in a wavelength band in which atmospheric water vapor absorbs and emits radiation.

Weather Research and Forecast (WRF) model. A community forecast model used in various configurations by universities, private-sector groups, and government agencies.

wet avalanches. Avalanches involving wet snow in which melting and water movement weaken the snowpack.

wild snow. New snow with a water content of 4 percent or less.

wind slabs. Cohesive snow layers created during wind transport.

windward. Facing the prevailing wind.

Wyssen avalanche towers. Permanently mounted towers that include an array of explosive charges that can be remotely controlled to lower and detonate over the snowpack to trigger avalanches.

Yamayuki. A snowstorm generated by the Sea of Japan with the largest accumulations in the mountains.

Bibliography

RECOMMENDED READING

Atwater, Montgomery M. 1968. *The Avalanche Hunters*. Philadelphia: Macrea Smith.

Bradley, Tyson. 2015. *Backcountry Skiing Utah: A Guide to the State's Best Ski Tours*. 3rd ed. Guilford, CT: Falcon Guides.

Burroughs, William. 1996. *Mountain Weather: A Guide for Skiers and Hillwalkers*. Ramsbury, UK: Crowood.

Burt, Christopher C. 2007. *Extreme Weather: A Guide and Record Book*. New York: Norton.

Dawson, Louis W. 1997. *Wild Snow: A Historical Guide to North American Ski Mountaineering*. Golden, CO: American Alpine Club.

Doesken, Nolan. J., and Arthur Judson. 1997. *The Snow Booklet: A Guide to the Science, Climatology, and Measurement of Snow in the United States*. Fort Collins: Colorado Climate Center.

Engen, Alan K., and Gregory C. Thompson. 2001. *First Tracks: A Century of Skiing in Utah*. Layton, UT: Gibbs Smith.

Ferguson, Sue A., and Edward R. LaChapelle. 2003. *The ABCs of Avalanche Safety*. 3rd ed. Seattle: Mountaineers Books.

Fredston, Jill, and Doug Fesler. 2011. *Snow Sense: A Guide to Evaluating Avalanche Hazard*. 5th ed. Edited by K. Birkeland and D. Chabot. Anchorage: Alaska Mountain Safety Center.

Garber, Howie. 2012. *Utah's Wasatch Range: Four Season Refuge*. 2nd ed. Johnson City, TN: Mountain Trail. https://www.howiegarberimages.com/award-winning-photography-books.

Gladwell, Malcolm. 2005. *Blink: The Power of Thinking without Thinking*. New York: Little, Brown.

Kelner, Alexis. 1980. *Skiing in Utah: A History*. Salt Lake City: Alexis Kelner.

LaChapelle, Edward R. 1969. *Field Guide to Snow Crystals*. Cambridge: International Glaciological Society.

LaChapelle, Edward R. 2001. *Secrets of the Snow: Visual Clues to Avalanche and Ski Conditions*. Seattle: University of Washington Press.

Libbrecht, Kenneth. 2006. *Ken Libbrecht's Field Guide to Snowflakes*. St. Paul: Voyageur.

Libbrecht, Kenneth. 2009. *The Secret Life of a Snowflake: An Up-Close Look at the Art and Science of Snowflakes*. Minneapolis: Voyageur.

Libbrecht, Kenneth, and Patricia Rasmussen. 2003. *The Snowflake*. Stillwater, MN: Voyageur.

Mass, Cliff. 2021. *The Weather of the Pacific Northwest*. 2nd ed. Seattle: University of Washington Press.

McClung, David, and Peter Schaerer. 2006. *The Avalanche Handbook*. 3rd ed. Seattle: Mountaineers Books.

McLean, Andrew. 1998. *The Chuting Gallery: A Guide to Steep Skiing in the Wasatch Mountains*. Park City: Paw Prince.

Renner, Jeff. 2005. *Mountain Weather*. Seattle: Mountaineers Books.

Tremper, Bruce. 2018. *Staying Alive in Avalanche Terrain*. 3rd ed. Seattle: Mountaineers Books.

Whiteman, C. David. 2000. *Mountain Meteorology: Fundamentals and Applications*. New York: Oxford University Press.

WEB RESOURCES

American Avalanche Association, httul://www.avalanche.org.
Canadian Avalanche Center, http://www.avalanche.ca.
Cottonwood Canyons Forecasts, https://www.weather.gov/slc/mtnwx?Cottonwood.
European Avalanche Warning Services, http://www.avalanches.org.
Know Before You Go (Avalanche Awareness Program), https://kbyg.org/.
MesoWest Weather Observations, http://mesowest.utah.edu.
National Weather Service Salt Lake City Forecast Office, https://www.weather.gov/slc/.
NCAR/RAL Real-Time Weather Data, http://weather.ral.ucar.edu.
New Zealand Avalanche Centre, http://www.avalanche.net.nz.
Opensnow, http://opensnow.com.
Penn State E-Wall, http://www.meteo.psu.edu/fxg1/ewall.html.
Ski Utah, http://www.skiutah.com.
TwisterData, http://www.twisterdata.com.
University of Utah Weather Center, http://weather.utah.edu.
Utah Avalanche Center, http://utahavalanchecenter.org.
Wasatch Snow Info, http://wasatchsnowinfo.com/Wasatch.
Wasatch Weather Weenies (Jim's blog), http://wasatchweatherweenies.blogspot.com.

CITED REFERENCES

Alcott, Trevor I., W. James Steenburgh, and Neil F. Laird. 2012. "Great Salt Lake-Effect Precipitation: Observed Frequency, Characteristics, and Associated Environmental Factors." *Weather and Forecasting* 27(4): 954–971. http://dx.doi.org/10.1175/WAF-D-12-00016.1.

Alta Powder News. 2009. 134 (Early Winter). http://centralpt.com/upload/400/10429_AltaSpring2009HistoricalPowderNews.pdf.

Baxter, Martin A., Charles E. Graves, and James T. Moore. 2005. "A Climatology of Snow-to-Liquid Ratio for the Contiguous United States." *Weather and Forecasting* 20(5): 729–744. http://dx.doi.org/10.1175/WAF856.1.

Cox, Justin A. W., W. James Steenburgh, David E. Kingsmill, Jason C. Shafer, Brian A. Colle, Olivier Bousquet, Bradley F. Small, and Huaqing Cai. 2005. "The Kinematic Structure of a Wasatch Mountain Winter Storm during IPEX IOP3." *Monthly Weather Review* 133(3): 521–542. http://dx.doi.org/10.1175/MWR-2875.1.

Dawson, Louis W. 1997. *Wild Snow: A Historical Guide to North American Ski Mountaineering.* Golden, CO: American Alpine Club.

Fetterer, Florence, Kenneth Knowles, Walter N. Meier, Matthew Savoie, and Ann K. Windnagel. 2017, updated daily. Sea Ice Index, Version 3 [Sea-Ice Extent]. Boulder, Colorado USA. NSIDC: National Snow and Ice Data Center. https://doi.org/10.7265/N5K072F8. Dataset accessed 19 April 2022.

Fischer, Erich, and Mathias Hauser. 2021. Summary for Policymakers of the Working Group I Contribution to the IPCC Sixth Assessment Report—data for Figure SPM.5 (v20210809). NERC EDS Centre for Environmental Data Analysis, 09 August 2021. http://dx.doi.org/10.5285/2787230b963942009e452255a3880609.

GISTEMP Team. 2022. *GISS Surface Temperature Analysis (GISTEMP), version 4.* NASA Goddard Institute for Space Studies. Dataset accessed 14 April 2022 at https://data.giss.nasa.gov/gistemp/.

Gillett, Nathan P., Elizaveta Malinina, Darrell Kaufman, and Raphael Neukom. 2021. "Summary for Policymakers of the Working Group I Contribution to the IPCC Sixth Assessment Report—data for Figure SPM.1 (v20210809)." NERC EDS Centre for Environmental Data Analysis. http://dx.doi.org/10.5285/76cad0b4f6f141ada1c44a4ce9e7d4bd.

Greene, Ethan, Thomas Wiesinger, Karl Birkeland, Cécile Coléou, Alan Jones, and Grant Statham. 2006. "Fatal Avalanche Accidents and Forecasted Danger Levels: Patterns in the United States, Canada, Switzerland, and France." In *Proceedings, 2006 International Snow Science Workshop.* Telluride, CO.

Horel, John D., Lloyd R. Staley, and Timothy W. Barker. 1988. "The University of Utah Interactive Dynamics Program—One Approach to Interactive Access and Storage of Meteorological Data." *Bulletin of the American Meteorological Society* 69(11): 1321–1327. http://dx.doi.org/10.1175/1520–0477(1988)069<1321:TUOUID>2.0.CO;2.

IPCC. 2021. "Summary for Policymakers." In *Climate Change 2021: The Physical Science Basis. Contribution of Working Group I to the Sixth Assessment Report of the Intergovernmental Panel on Climate Change*, edited by Valérie Masson-Delmotte, Panmao Zhai, Anna Pirani, Sarah L. Connors, Clotilde Péan, Sophie Berger, Nada Caud, Yang Chen, Leah Goldfarb, Melissa I. Gomis, Mengtian Huang, Katherine Leitzell, Elisabeth Lonnoy, J. B. Robin Matthews, Tom K. Maycock, Thomas Waterfield, Özge Yelekçi, Rong Yu and Baiquan Zhou. Cambridge University Press. In Press.

Isotta, Francesco A., et al. 2013. "The Climate of Daily Precipitation in the Alps: Development and Analysis of a High-Resolution Grid Dataset from Pan-Alpine Rain-Gauge Data." *International Journal of Climatology* 34(5): 1657–1675.

Jones, Leigh P. 2010. "Assessing the Sensitivity of Wasatch Snowfall to Temperature Variations." Master's thesis, University of Utah.

Judson, Arthur, and Nolan Doesken. 2000. "Density of Freshly Fallen Snow in the Central Rocky Mountains." *Bulletin of the American Meteorological Society* 81(7): 1577–1587. http://dx.doi.org/10.1175/1520-0477(2000)081<1577:DOFFSI>2.3.CO;2.

Kalitowski, Mark. 1988. "The Avalanche History of Alta." *Avalanche Review* 7(3), 3.

Karl, Thomas R., and Kevin E. Trenberth. 2003. "Modern Global Climate Change." *Science* 302(5651): 1719–1723. http://dx.doi.org/10.1126/science.1090228.

Kelner, Alexis. 1980. *Skiing in Utah: A History*. Salt Lake City: Alexis Kelner.

LaChapelle, Edward R. 1962. *The Density Distribution of New Snow*. Project F, Progress Rep. 2, USDA Forest Service. Salt Lake City: Wasatch National Forest, Alta Avalanche Center.

Lackner, Christian P., Bart Geerts, and Yonggang Wang. 2021. "Impact of Global Warming on Snow in Ski Areas: A Case Study using a Regional Climate Simulation over the Interior Western United States." *Journal of Applied Meteorology and Climatology* 60(5): 677–694. https://journals.ametsoc.org/view/journals/apme/60/5/JAMC-D-20-0155.1.xml.

Lenssen, Nathan, Gavin Schmidt, James Hansen, Matthew Menne, Avraham Persin, Reto Ruedy, and Daniel Zyss. 2019. "Improvements in the GISTEMP Uncertainty Model." *Journal of Geophysical Research Atmospheres* 124(12): 6307–6326. https://agupubs.onlinelibrary.wiley.com/doi/10.1029/2018JD029522.

Neiman, Paul J., F. Martin Ralph, Allen B. White, David E. Kingsmill, and P. Ola G. Persson. 2002. "The Statistical Relationship between Upslope Flow and Rainfall in California's Coastal Mountains: Observations during CALJET." *Monthly Weather Review* 130(6): 1468–1492. http://dx.doi.org/10.1175/1520-0493(2002)130<1468:TSRBUF>2.0.CO;2.

Olefs, Mark, Herbert Formayer, Andreas Gobiet, Thomas Marke, Wolfgang Schöner, and Michael Revesz. 2021. "Past and Future Changes of the Austrian Climate—Importance for Tourism." *Journal of Outdoor Recreation and Tourism* 34: 1–13. https://doi.org/10.1016/j.jort.2021.100395.

Snellman, Leonard W. 1982. "Impact of AFOS on Operational Forecasting." In *Preprints, 9th Conference on Weather Forecasting and Analysis*, 13–16. Seattle, WA: American Meteorological Society.

Steenburgh, W. James, and Trevor I. Alcott. 2008. "Secrets of the Greatest Snow on Earth." *Bulletin of the American Meteorological Society* 89(9): 1285–1293. http://dx.doi.org/10.1175/2008BAMS2576.1.

Tarboton, David. 2017. "Great Salt Lake Bathymetry." HydroShare. http://www.hydroshare.org/resource/582060f00f6b443bb26e896426d9f62a.

WGMS. 2021. "Fluctuations of Glaciers Database." World Glacier Monitoring Service. Dataset accessed 18 April 2022 at http://dx.doi.org/10.5904/wgms-fog-2021–05.

Wiese, David N., D.-N. Yuan, Carmen Boening, Felix W. Landerer, and Michael M. Watkins. 2019. JPL GRACE and GRACE-FO Mascon Ocean, Ice, and Hydrology Equivalent Water Height RL06M CRI Filtered Version 2.0, PO.DAAC, CA, USA. Dataset accessed 19 April 2022 at http://dx.doi.org/10.5067/TEMSC-3MJ62.

Yeager, Kristen N., W. James Steenburgh, and Trevor I. Alcott. 2013. "Contributions of Lake-Effect Periods to the Cool-Season Hydroclimate of the Great Salt Lake Basin." *Journal of Applied Meteorology and Climatology* 52(2): 341–362. http://dx.doi.org/10.1175/JAMC-D-12–077.1.

Zemp, Michael, et al. 2019. "Global Glacier Mass Changes and Their Contributions to Sea-Level Rise from 1961 to 2016." *Nature* (568): 382–386. https://doi.org/10.1038/s41586-019-1071-0.

Index

Page numbers followed by *f* indicate figures.

About the Author

Dr. Jim Steenburgh is a professor of atmospheric sciences at the University of Utah, an avid backcountry and resort skier, and creator of the popular blog Wasatch Weather Weenies. An award-winning teacher and leading authority on the weather and climate of the Wasatch Mountains and western United States, his research on snow, winter storms, and forecasting has been featured by *The Weather Channel*, the *New York Times, USA Today*, and the *Salt Lake Tribune*.